London Mathematical Society
Lecture Note Series 7

Introduction to Combinatory Logic

J. R. HINDLEY
B. LERCHER and
J. P. SELDIN

D0146501

CAMBRIDGE UNIVERSITY PRESS

London Mathematical Society Lecture Note Series 7

Introduction to Combinatory Logic

J. R. HINDLEY, B. LERCHER
and J. P. SELDIN

CAMBRIDGE AT THE UNIVERSITY PRESS 1972

WILLIAM MADISON RANDALL LIBRARY UNC AT WILMINGTON

Published by The Syndics of the Cambridge University Press
Bentley House, 200 Euston Road, London N. W. 1.
American Branch: 32 East 57th Street, New York, N. Y. 10022

© Cambridge University Press 1972

Library of Congress Catalogue Card No.: 74-179165

ISBN: 0 521 09697 9

Printed in Great Britain
at the University Printing House, Cambridge
(Brooke Crutchley, University Printer)

Contents

QA9
.H525

105927

Introduction

Combinatory logic deals with a class of formal systems designed for studying certain primitive ways in which functions can be combined to form other functions.

These notes present some of the basic techniques and results in the subject, as well as two or three more special topics. There is no attempt to be comprehensive, but merely to give the flavour of the subject. Most long proofs will be omitted, and replaced by references to the published literature.

The reader is assumed to have no previous knowledge of combinatory logic, but to have some experience of predicate calculus and recursive functions.

The subject will be introduced via λ-conversion, since this system has approximately the same purpose as combinatory logic and is intuitively a little easier to understand. After the introduction of combinators in Chapter 2, four more or less independent topics will be covered: Chapters 3 and 4 will deal with recursive functions, Chapters 5, 6 and 7 with the extensional theory of combinators, Chapters 9 and 10 with combinator-based systems of logic, and Chapters 8 and 11 with a proof-theoretic application.

This choice of topics reflects merely the current interests of the authors, not their beliefs as to what is most important. A growing topic from which several interesting new problems are arising is the use of λ-conversion and combinators in the theory of computing languages. Unfortunately, due to the authors' ignorance, this cannot be dealt with here, but it is hoped that the earlier chapters will be of some interest to people in this field. (For some references to this topic, see the end of Chapter 2.)

Two other topics which have had to be omitted are logical type-theories and the logical systems studied by F. B. Fitch. References for these will be given in Chapter 10.

Simple exercises have been added to the earlier chapters to give manipulative practice. Suitable references for further study are listed in the bibliography. These are referred to in the text by the author's name followed by an abbreviation of the title, e. g. 'Kleene [IMM]'. The fullest current treatment of the subject is in Curry and Feys Combinatory Logic volume I [CLg. I], and its forthcoming sequel [CLg. II].

These notes are based on lectures given at the Universities of Bristol (1967-8), Swansea (1968-9), and Southern Illinois (1969-70). Very little of the material is original, the most important sources being [CLg. I] and [CLg. II]. The authors wish to thank those who listened to the lectures for their interest and helpful comments. We also thank H. B. Curry for his permission to use some results from the manuscript of [CLg. II] before publication, P. Martin-Löf and W. W. Tait for permission to use their forthcoming proof of the Church-Rosser theorem, K. Schütte for part of the proof of Theorem 11. 17, and Miss J. Görtz for typing most of the manuscript. Finally, we acknowledge our debt to H. B. Curry for teaching us most of what we know about combinators.

1. Lambda-conversion

As mentioned in the introduction, the theory of λ-conversion is concerned mainly with functions. A function f associates at most one object $f(x_1, \ldots, x_n)$, its <u>value</u>, with each n-tuple of objects $x_1, \ldots, x_n$, its <u>arguments</u>, which might themselves be functions in the present context.

In everyday differential calculus, an expression such as 'x - y' can be considered as defining either a function f of x or a function g of y. A convenient way to distinguish these two functions is to introduce a symbol 'λ' and define

$$f = \lambda x. x - y, \qquad g = \lambda y. x - y .$$

We say that prefixing 'λx' <u>abstracts</u> the function $\lambda x. x - y$ from the expression 'x - y'. This gives us a systematic way of constructing, for each expression involving 'x', a notation for the corresponding function of x. This is the starting-point of the theory of λ-conversion.

For the above function f, we have for example,

$$f(0) = 0 - y, \qquad f(1) = 1 - y .$$

In the λ-notation these equations become

$$(\lambda x. x - y)(0) = 0 - y, \qquad (\lambda x. x - y)(1) = 1 - y .$$

This notation can be extended to functions of more than one variable. For example, to the expression 'x - y' correspond two functions h, k of two variables, defined by

$$h(x, y) = x - y, \qquad k(y, x) = x - y .$$

These can be denoted by

$$h = \lambda xy.\, x - y\ , \qquad k = \lambda yx.\, x - y\ .$$

However, we can avoid the need for a special notation for functions of several variables by using functions whose values are not numbers but functions. For example, instead of using the two-place function h above, we could represent h by the function h* defined by

$$h^* = \lambda x.\,(\lambda y.\, x - y)\ .$$

For each number a,

$$h^*(a) = \lambda y.\, a - y\ ;$$

hence for each pair a, b,

$$(h^*(a))(b) = (\lambda y.\, a - y)(b) = a - b\ .$$

For this reason we shall only need a λ-notation for functions of one variable.

Let us now look at a formal system of λ-conversion which embodies the above ideas.

We begin by assuming that there is an infinite sequence of <u>variables</u> and a finite or infinite sequence of <u>constants</u>. An <u>atom</u> is a variable or a constant.

Definition 1.1 (λ-<u>terms</u>). The set of λ-terms is defined by induction as follows:

(i) Every atom is a λ-term.

(ii) If X and Y are λ-terms, then (XY) is a λ-term.

(iii) If Y is a λ-term and x is a variable, then (λx. Y) is a λ-term.

<u>Examples of</u> λ-<u>terms</u>

$(\lambda x.\,(xy))$, $((\lambda y.\, y)(\lambda x.\,(xy)))$,

$(x(\lambda x.\,(\lambda x.\, x)))$, $(\lambda x.\, y)$.

Notation. Letters 'x', 'y', 'z', 'u', 'v', 'w' will denote variables, and distinct letters will denote distinct variables unless stated otherwise. Capital Roman letters will denote arbitrary λ-terms. Parentheses will be omitted in such a way that, for example, 'WXYZ' denotes the λ-term (((WX)Y)Z) and 'λx. XY' denotes (λx. (XY)). Identity of terms, numbers, etc. will be denoted by '≡' ; the symbol '=' will be used for another relation later on. We shall shorten 'if and only if' to 'iff'. Finally, we shall use the abbreviation

$$\lambda x_1 \ldots x_n . Y \equiv (\lambda x_1 . (\lambda x_2 . (\ldots (\lambda x_n . Y) \ldots))) .$$

It is assumed of course that the three classes of terms (atoms, λx. Y, XY) do not intersect; also that if $XY \equiv UV$ then $X \equiv U$ and $Y \equiv V$, and if $\lambda x. Y \equiv \lambda u. V$ then $x \equiv u$ and $Y \equiv V$.

Interpretation. The exact interpretation of the λ-terms varies in different applications of the theory. In general each λ-term is intended to represent a one-place function, whose values and arguments might themselves be functions. The variables represent arbitrary (one-place) functions, and (XY) represents the result of applying the function X to the argument Y. If Y is a term in which x occurs free (defined below), then (λx. Y) represents the function whose value at an argument A is the result of substituting A for x in Y. For example, λx. xy represents the operation of applying a function to a particular argument y, so that for all terms F,

$$(\lambda x. xy)F = Fy$$

holds in the sense that both sides of the equation have the same interpretation. Again, λx. y represents the constant function which takes the value y for all arguments, so that

$$(\lambda x. y)F = y$$

holds in the same sense.

Notice that a term (XX) represents the result of applying a function to itself; this sort of term could be excluded by type-restrictions, and in applications of the theory these terms are usually left uninterpreted. But admitting self-applying functions into mathematics does not necessarily lead to contradictions, and furthermore, one of the original aims of combinatory logic was to study the most primitive properties of the function concept with as few restrictions as possible (see Remark 2.23 later). Hence the terms will be left unrestricted for the present.

Definition 1.2 (Occurrence). The relation X occurs in Y (X is part of Y, or Y contains X) is defined by induction on the construction of Y as follows:

(i) X occurs in X.

(ii) If X occurs in U or in V, then X occurs in (UV).

(iii) If X occurs in U, then for every y, X occurs in $(\lambda y. U)$.

The meaning of 'an occurrence of X in Y' is assumed to be intuitively clear; for example, there are two occurrences of (xy) in $(xy)(\lambda x. xy)$. (The reader who wants more precision can define an occurrence of X in Y to be a particular tree-form deduction from (i), (ii), and (iii) of the fact that X occurs in Y.)

Note that by Definition 1.2, x does not occur in $(\lambda x. y)$.

Definition 1.3 (Free and bound variables). An occurrence of a variable x in Y is bound iff it is inside a part of Y of the form $\lambda x. Z$; otherwise it is free. We say that x is free in Y (or '$x \in Y$') just when x has a free occurrence in Y.

Note that x can be free and bound in the same term; for example, in $(x(\lambda x. x))$.

Definition 1.4 (Substitution). For any terms N, M, and any variable x, the result [N/x]M of substituting N for every free occurrence of x in M (and changing bound variables to avoid clashes) is defined as follows (after Curry and Feys [CLg. I], p. 94) by induction on the construction of M:

(i) $[N/x]x \equiv N$.

(ii) $[N/x]a \equiv a$ for all atoms $a \not\equiv x$.

(iii) $[N/x](M_1 M_2) \equiv ([N/x]M_1)([N/x)M_2)$.

(iv) $[N/x](\lambda x.\, Y) \equiv \lambda x.\, Y$.

(v) $[N/x](\lambda y.\, Y) \equiv (\lambda y.\, [N/x]Y)$ if $y \not\equiv x$, and $y \notin N$ or $x \notin Y$,

$\equiv (\lambda z.\, [N/x][z/y]Y)$ if $y \not\equiv x$ and $y \in N$ and $x \in Y$.

Here z is the first variable free in neither N nor Y.

The last clause prevents the intuitive meaning of $[N/x](\lambda y.\, Y)$ from depending on the bound variable y. For example, without this clause we would have, for all y and w,

$$[w/x](\lambda y.\, x) \equiv \lambda y.\, w ,$$

which would represent a constant-function if $y \not\equiv w$, but the identity-function if $y \equiv w$. But with this clause included, when $y \equiv w$ we have

$$[w/x](\lambda w.\, x) \equiv \lambda z.\, w ,$$

which is still a constant-function. Actually Church, the originator of λ-conversion, simply left $[N/x]M$ undefined in the last part of case (v), but it seems tidier to define it for all N, x and M.

Definition 1.5. A *change of bound variables* in a term X is the replacement of a part of X with the form $\lambda y.\, Z$, with y not bound in Z, by $\lambda v.\, [v/y]Z$ for any v which is neither free nor bound in Z. X is *congruent* to Y iff Y is the result of applying a series of changes of bound variables to X.

Since the reverse of a change of bound variable is also a change of bound variable, congruence is symmetric; it is also reflexive and transitive. Congruent terms have identical interpretations and play identical roles in any application of λ-conversion, so they are usually regarded as identical.

Incidentally, the above definition of change of bound variables differs slightly from the corresponding definitions in Curry and

Feys [CLg. I] and Church [CLC], but the three congruence relations are the same.

Definition 1. 6 (Reduction; Curry's $\lambda\beta$-reduction). X contracts to Y iff Y is the result of replacing a part of X of the form $(\lambda x. M)N$ by $[N/x]M$. X reduces to Y (or $X \triangleright Y$) iff Y is obtained from X by a finite (perhaps empty) series of contractions and changes of bound variables. (That is, $\triangleright$ is the transitive and reflexive relation generated by contraction and changing bound variables.) Any term of form

$$(\lambda x. M)N$$

is called a redex, and $[N/x]M$ is called its contractum.

From an intuitive point of view, the function $\lambda x. M$ applied to N has the value $[N/x]M$, so that reduction represents, in a sense, the process of calculating the values of functions occurring in a term.

Examples:

1. $(\lambda x. xy)F \triangleright Fy$.
2. $(\lambda x. y)F \triangleright y$.
3. $(\lambda x. (\lambda y. yx)z)v \triangleright [v/x]((\lambda y. yx)z) \equiv (\lambda y. yv)z$
   $\triangleright [z/y](yv) \equiv zv$.
4. $(\lambda x. xxy)(\lambda x. xxy) \triangleright (\lambda x. xxy)(\lambda x. xxy)y$
   $\triangleright (\lambda x. xxy)(\lambda x. xxy)yy$
   etc.
5. $(\lambda x. xx)(\lambda x. xx) \triangleright (\lambda x. xx)(\lambda x. xx)$
   etc.

Definition 1. 7. A term X is said to be in normal form iff X contains no redexes. If a term U reduces to an X in normal form, then X is called a normal form of U.

Roughly speaking, a normal form of U is the simplest term with the same intuitive interpretation as U. For instance, in Example 3 above, zv is a normal form of $(\lambda x. (\lambda y. yx)z)v$.

The terms in Examples 4 and 5 have no normal forms.

Now, many terms can be reduced in more than one way; for instance, in Example 3 we have also

$$(\lambda x.\,(\lambda y.\,yx)z)v \rhd (\lambda x.\,zx)v \qquad \text{(contracting } (\lambda y.\,yx)z)$$
$$\rhd\ zv\,.$$

In this case both reductions reach the same normal form. Is this always true? The next theorem and its corollary show that it is if we ignore the difference between congruent terms.

Theorem 1.8 (Church-Rosser Theorem, first form). If $U \rhd X$ and $U \rhd Y$, then there exists a Z such that $X \rhd Z$ and $Y \rhd Z$.

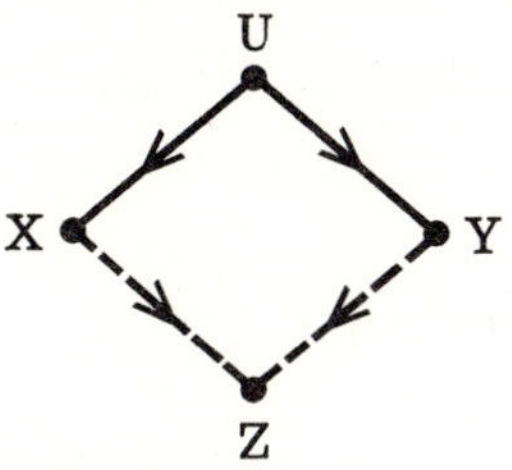

Proof. See Appendix 1.

Corollary 1.8.1. If U has normal forms X and Y, then X is congruent to Y.

Proof. Since $U \rhd X$ and $U \rhd Y$, the theorem gives us a Z such that $X \rhd Z$ and $Y \rhd Z$. Since X and Y contain no redexes, they must be congruent to Z. Hence, X is congruent to Y.

Let us now look at an alternative definition of reduction which is technically simpler than Definition 1.6, though perhaps not so easy to understand.

Definition 1.9 (Alternative definition of reduction).

Axiom-schemes:

(α) $\lambda y.\,Z \rhd \lambda v.\,[v/y]Z$ if y is not bound in Z and v is not free or bound in Z).

(β) $(\lambda x. M)N \rhd [N/x]M$.

(ρ) $M \rhd M$.

<u>Deduction-rules</u>:

(μ) $X \rhd X' \Rightarrow ZX \rhd ZX'$.

(ν) $X \rhd X' \Rightarrow XZ \rhd X'Z$.

(ξ) $X \rhd X' \Rightarrow \lambda x. X \rhd \lambda x. X'$.

(τ) $X \rhd Y$ and $Y \rhd Z \Rightarrow X \rhd Z$.

We say that $X \rhd Y$ iff there is a proof of this statement using only the above axioms (i. e. any particular cases of the axiom-schemes for particular terms Z, v, y, M, N, x) and the above deduction-rules.

Lemma 1. 10. <u>Definitions 1. 6 and 1. 9 define the same relation.</u>

Proof. Straightforward.

Definition 1. 11 (λ-<u>convertibility</u>, λ-<u>equality</u>). We say that X <u>is λ-convertible to</u> Y (or X <u>is equal to</u> Y, or $X = Y$) iff this statement can be deduced from the axiom schemes (α), (β), (ρ) in Definition 1. 9 (with '$\rhd$' replaced by '='), by the deduction-rules (μ), (ν), (ξ), (τ), and

(σ) $X = Y \Rightarrow Y = X$.

Lemma 1. 12. $X = Y$ <u>iff</u> Y <u>is obtained from</u> X <u>by a finite (perhaps empty) series of reductions and reversed reductions. (That is,</u> = <u>is the equivalence relation generated by</u> $\rhd$.)

The next theorem is an important one; among other things, it shows that if $X = Y$, then X and Y both intuitively represent the same function, since they can both be reduced to one term (this explains the notation '='). It is also our principal tool for proving terms unequal.

Theorem 1.13 *(Church-Rosser Theorem, second form).* *If* $X = Y$, *then there is a* Z *such that* $X \triangleright Z$ *and* $Y \triangleright Z$.

Proof. By induction on the number n of reductions and reversed reductions in Lemma 1.12. The basis, $n \equiv 1$, is easy. For the induction step we have

$$X = Y', \qquad Y' \triangleright Y \quad \text{or} \quad Y \triangleright Y';$$

and the induction hypothesis can be applied to $X = Y'$ to give a Z' such that

$$X \triangleright Z', \qquad Y' \triangleright Z'.$$

Then Theorem 1.8 applied to Y', Y and Z' gives the result.

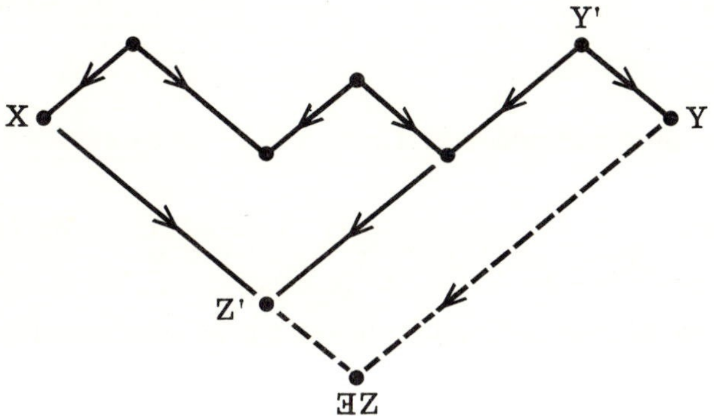

Corollary 1.13.1. *If* $X = Y$ *and* Y *is in normal form, then* $X \triangleright Y$.

Proof. By Theorem 1.13 there is a Z such that $X \triangleright Z$ and $Y \triangleright Z$. Since Y is in normal form, Z is congruent to Y. Hence $X \triangleright Y$.

Corollary 1.13.2. *If* $X = Y$, *then either* X *and* Y *do not have normal forms or* X *and* Y *both have the same normal forms.*

Corollary 1.13.3. Two equal terms in normal form must be congruent.

By this last corollary, $(\lambda x. x) \neq (\lambda x. xx)$, and hence the axioms and rules defining equality are consistent in the sense that not every two terms are equal.

Remark 1.14 (Extensionality). Equality of terms does not correspond to extensional equality of functions; there exist terms X and Y such that $XV = YV$ for all V, and yet $X \neq Y$. For example, if

$$X \equiv \lambda v. yv, \qquad Y \equiv y,$$

then for all V,

$$XV \triangleright yV \equiv YV.$$

To remedy this we could add to Definition 1.11 an extensionality rule, viz.

(ext) If $XV = YV$ for all V, then $X = Y$.

This rule will be studied in Chapter 5. But a lot can be done without it, as we shall see.

Remark 1.15 (Church's restricted λ-terms). If U has a normal form, not every part of U need have a normal form. For example

$$U \equiv (\lambda x. y)F,$$

where F has no normal form. In Church's original system of λ-conversion such cases were excluded by only allowing $\lambda x. Y$ to be a term when $x \in Y$. The reason for this was that terms without normal forms had no interpretation given to them, and Church did not want meaningful terms to contain meaningless parts. All the preceding results remain true for Church's restricted notion of term, but this will not always hold for future proofs. (The system we have been looking at here is what Church calls 'λ-K-conversion', and what Curry calls 'λβK-conversion'.)

For more information about λ-conversion see Church [CLC], or Curry and Feys [CLg. I] Chapters 3 and 4. Also, much of what follows could be written out in terms of λ-conversion instead of combinators.

Exercises

1. Express the following in an informal λ-notation, using 'D' for the differentiation operator.

(a) The derivative of x^2 is $2x$.

(b) The derivative of x^2 at the point 3 is 6.

(c) The derivative at x of the function g defined by $g(x) = f(x^2)$ (for a given f) is distinct from the derivative at x^2 of f.

2. Evaluate

$$[(\lambda y.\, xy)/x](\lambda y.\, x(\lambda x.\, x)) \,, \qquad [(\lambda y.\, vy)/x](y(\lambda v.\, xv)) \,.$$

3. Prove that the relation of congruence (Definition 1.5) is symmetric.

4. Find a term X which is not in normal form, but which is irreducible; i.e. which satisfies

$$X \triangleright Y \Rightarrow Y \equiv X \,.$$

(In fact, modulo congruence, there is only one such X with the form $(\lambda x.\, M)N$.)

5. Prove that if we add to Definition 1.11 an axiom-scheme

$$\lambda x.\, Mx = M \qquad (x \notin M) \,,$$

then the extensionality rule in Remark 1.14 is valid for the resulting equality. (Cf. Theorem 5.4.)

2. Combinators

Systems of combinators are designed to perform the same tasks as systems of λ-conversion, but manage to carry them out without using bound variables. Thus the annoying technical complications concerned with substitution and congruence will not be needed here at all. However, for this technical advantage we shall have to sacrifice the intuitive simplicity of the λ-notation.

In arithmetic, the commutative law of addition can be expressed as

'for all x, y; $x + y = y + x$.'

But this law can be expressed without using the bound variables 'x' and 'y' by defining

$A(x, y) = x + y$ for all x, y,

and by introducing an operator **C** defined by

$(\mathbf{C}(f))(x, y) = f(y, x)$ for all f, x, y.

Then the law becomes simply

$\mathbf{C}A = A$.

The operator **C** may be called a combinator; other examples of combinators are the following.

B, which operates on 2 functions $(\mathbf{B}(f, g))(x) = f(g(x))$.

I, the identity operator, $\mathbf{I}(f) = f$.

K, which forms constant functions, $(\mathbf{K}(a))(x) = a$.

S, defined by the property $(\mathbf{S}(f, g))(x) = f(x, g(x))$.

Instead of trying to define 'combinator' rigorously in this informal context, let us build up a formal system of terms in which the above 'combinators' can be represented. The interpretation of these terms will be considered later, in Remark 2.23. (For further informal motivation, see the very clear exposition in Schönfinkel [BML].)

We begin by assuming that there is an infinite sequence of <u>variables</u>, and a finite or infinite sequence of <u>constants</u>, including three called <u>basic combinators</u>, **I**, **K**, **S**. These variables and constants are called <u>atoms</u>.

Definition 2.1 <u>(Combinatory terms).</u> The set of combinatory terms is defined by induction as follows:

(i) Every atom is a combinatory term.

(ii) If X and Y are combinatory terms, then (XY) is a combinatory term.

A <u>combinator</u> is a term whose only atoms are **I**, **K**, and **S**.

It is assumed that the set of atoms does not intersect the set of terms of form (XY), and that if $(XY) \equiv (UV)$ then $X \equiv U$ and $Y \equiv V$.

Notation. The conventions about variables, identity, and omitting parentheses will be the same as for λ-terms. But in this chapter capital Roman letters will denote combinatory terms, not λ-terms, and the word 'term' will mean 'combinatory term' unless otherwise stated.

The notion of <u>occurrence</u> is the same here as in Definition 1.2, except that now clause (iii) of that definition is irrelevant. There are no bound occurrences of variables in combinatory terms. The phrase 'x occurs in M' may be written '$x \in M$'.

Definition 2.2 <u>(Substitution).</u> For any N, M, and x, the term

$$[N/x]M$$

is defined as in Definition 1.4 except that clauses (iv) and (v) are now irrelevant. For mutually distinct $x_1, \ldots, x_n$ and for any terms $N_1, \ldots, N_n$, we similarly define

$$[N_1/x_1, \ldots, N_n/x_n]M$$

to be the result of simultaneously substituting N_1 for x_1, N_2 for x_2, ..., N_n for x_n, in M.

To make combinatory terms do the same work as λ-terms, we must be able to define abstraction of functions within the system, that is, we must define a reduction ▷ of combinatory terms, and for each variable x define an operation of abstraction (here written '[x]', and corresponding to λx), such that

$$([x]M)N \triangleright [N/x]M .$$

Note that [x] will not itself be part of the formalism as λx was, but will be an informal operator on terms, which acts on a term M to produce another term which we shall call '[x]M'. The operation will be defined by the clauses of Definition 2.10 below.

Definition 2.3 (Weak reduction). The relation X ▷ Y (X reduces to Y) is defined by induction as follows.

Axiom-schemes:

(**I**)	**I**X ▷ X	for all X.
(**K**)	**K**XY ▷ X	for all X, Y.
(**S**)	**S**XYZ ▷ XZ(YZ)	for all X, Y, Z.
(ρ)	X ▷ X	for all X.

Deduction-rules:

(μ) X ▷ X' ⇒ ZX ▷ ZX'.

(ν) X ▷ X' ⇒ XZ ▷ X'Z.

(τ) X ▷ Y and Y ▷ Z ⇒ X ▷ Z.

Any term **I**X, **K**XY, or **S**XYZ is called a redex, and the corresponding X, X, or XZ(YZ) (respectively) is called its contractum. Contracting a term X means replacing one occurrence of a redex in X by its contractum. X is said to be <u>in normal form</u> iff X contains no redexes. If U reduces to an X in normal form, this X is called <u>a normal form of</u> U.

Lemma 2.4. X ▷ Y iff Y is obtained from X by a finite (perhaps empty) series of contractions.

Example 2.5. Define **B** ≡ **S**(**KS**)**K**.
Then **B**XYZ ▷ X(YZ), because

BXYZ ≡ **S**(**KS**)**K**XYZ		
	▷ **KS**X(**K**X)YZ	by contracting **S**(**KS**)**K**X,
	▷ **S**(**K**X)YZ	by contracting **KS**X,
	▷ **K**XZ(YZ)	by contracting **S**(**K**X)YZ,
	▷ X(YZ)	by contracting **K**XZ.

Example 2.6. Define **C** ≡ **S**(**BBS**)(**KK**).
Then **C**XYZ ▷ XZY, because

CXYZ ≡ **S**(**BBS**)(**KK**)XYZ		
	▷ **BBS**X(**KK**X)YZ	by contracting **S**(**BBS**)(**KK**)X,
	▷ **BBS**X**K**YZ	by contracting **KK**X,
	▷ **B**(**S**X)**K**YZ	by Exercise 2.5,
	▷ **S**X(**K**Y)Z	by Exercise 2.5,
	▷ XZ(**K**YZ)	
	▷ XZY.	

(Incidentally, in line 4 of this reduction, a redex **K**YZ seems to occur; but this is not really so, since **B**(**S**X)**K**YZ is really ((((**B**(**S**X))**K**)Y)Z) when all the parentheses are put in.)

Example 2.7. **SKK**X ▷ **K**X(**K**X)
▷ X.

Hence we could have defined **I** ≡ **SKK** instead of taking **I** as an atom, but this would make the theory of strong (extensional) reduction in Chapter 7 untidy; see Remark 7.21.

Lemma 2. 8. For all $N_1, \ldots, N_m$ and mutually distinct $x_1, \ldots, x_m$, if $X \rhd Y$ then

$$[N_1/x_1, \ldots, N_m/x_m]X \rhd [N_1/x_1, \ldots, N_m/x_m]Y .$$

Proof. Use induction on the proof that $X \rhd Y$. Every substitution instance of an axiom in Definition 2. 3 is an axiom, and the same is true for the rules.

Lemma 2. 9. If $X \rhd Y$, then Y contains no variables that do not occur in X.

Definition 2. 10 (Abstraction). For each term M and variable x, a term $[x]M$ is defined by induction on the construction of M as follows:

(i) $[x]x \equiv \mathbf{I}$.

(ii) $[x]M \equiv \mathbf{K}M$ if $x \notin M$.

(iii) $[x]Ux \equiv U$ if $x \notin U$.

(iv) $[x]UV \equiv \mathbf{S}([x]U)([x]V)$ if neither (ii) nor (iii) applies.

Example 2. 11.

$[x]xy \equiv \mathbf{S}([x]x)([x]y)$ by (iv),

$\equiv \mathbf{SI}(\mathbf{K}y)$ by (i) and (ii).

Theorem 2. 12. (Compare (β) in Definition 1. 9.) For all M, N and x,

$$([x]M)N \rhd [N/x]M .$$

Proof. By Lemma 2. 8 it is enough to prove that

$$([x]M)x \rhd M .$$

This will be done by induction on the construction of M. There are the following cases:

Case 1: $M \equiv x$. Then, using Definition 2. 10(i),

$([x]x)x \equiv \mathbf{I}x \rhd x$.

Case 2: M is an atom distinct from x. Then $x \notin M$, and hence, by (ii),

$$([x]M)x \equiv \mathbf{K}Mx \rhd M .$$

Case 3: $M \equiv UV$. Then by the induction hypothesis, we have

$$([x]U)x \rhd U , \qquad ([x]V)x \rhd V .$$

Subcase a: $x \notin M$. Then the proof is like Case 2 above.

Subcase b: $x \notin U$ and $V \equiv x$. Then, using Definition 2.10(iii),

$$\begin{aligned} ([x]M)x &\equiv ([x]Ux)x \\ &\equiv Ux \equiv M . \end{aligned}$$

Subcase c: Neither of the two above subcases applies. Then

$$\begin{aligned} ([x]M)x &\equiv \mathbf{S}([x]U)([x]V)x && \text{by (iv)} \\ &\rhd ([x]U)x(([x]V)x) \\ &\rhd UV && \text{by the induction hypothesis} \\ &\equiv M . \end{aligned}$$

Note how the reduction axioms for **I**, **K**, and **S** fit in with the cases in this proof; this is their purpose.

Incidentally there are several other possible definitions of $[x]M$ besides this one. For example, clause (ii) could be restricted to the case in which M is an atom, and clause (iii) could be restricted or omitted, and Theorem 2.12 would still be true. However, if these modifications were made, the useful Lemma 2.17(b) and Theorem 6.3(c) would fail.

Definition 2.13. For variables $x_1, \ldots, x_m$ (not necessarily distinct), define

$$[x_1, \ldots, x_m]M \equiv [x_1]([x_2](\ldots([x_m]M)\ldots)) .$$

Example 2.14. $[x, y]x \equiv [x]([y]x)$

$$\begin{aligned} &\equiv [x](\mathbf{K}x) && \text{by Definition 2.10(ii)} \\ &\equiv \mathbf{K} && \text{by (iii)} . \end{aligned}$$

Example 2.15. $[x, y, z]xz(yz) \equiv [x]([y]([z]xz(yz)))$

$$\equiv [x]([y](\mathbf{S}([z]xz)([z]yz))) \quad \text{by (iv)},$$
$$\equiv [x]([y]\mathbf{S}xy) \quad \text{by (iii)},$$
$$\equiv [x]\mathbf{S}x \quad \text{by (iii)},$$
$$\equiv \mathbf{S} \quad \text{by (iii)}.$$

Theorem 2.16. For distinct variables $x_1, \ldots, x_m$,

$$([x_1, \ldots, x_m]M) N_1 \ldots N_m \;\triangleright\; [N_1/x_1, \ldots, N_m/x_m]M .$$

Proof. By Lemma 2.8 it is enough to show that

$$([x_1, \ldots, x_m]M) x_1 \ldots x_m \;\triangleright\; M ,$$

which can easily be proved by induction on m, using Theorem 2.12.

Lemma 2.17. (a) If $v \notin Z$, then $[v][v/y]Z \equiv [y]Z$.

(b) If $y \not\equiv x$ and $y \notin N$, then

$$[N/x][y]Z \equiv [y][N/x]Z .$$

Proofs. By induction on the construction of Z. Part (b) depends on the fact that y occurs in $[N/x]Z$ iff y occurs in Z.

This lemma will be needed in Chapter 3. Part (a) compares with the comments on changing bound variables after Definition 1.5, and (b) compares with Definition 1.4(iv) on substitution in λ-terms.

The following theorem corresponds to Theorem 1.8.

Theorem 2.18 (Rosser). If $U \triangleright X$ and $U \triangleright Y$, then there exists a Z such that $X \triangleright Z$ and $Y \triangleright Z$.

Proof. See the remark at the end of Appendix 1. This proof is much simpler than for λ-reduction because combinatory contractions are simpler.

Corollary 2.18.1. A combinatory term can have at most one normal form. (Cf. Corollary 1.8.1.)

Definition 2.19 (Weak Equality). We say that $X = Y$ holds iff this statement can be deduced from the axiom-schemes **(I)**, **(K)**, **(S)**, (ρ) and the rules (μ), (ν), (τ) of Definition 2.3 (with '=' instead of '$\triangleright$'), together with the rule

$$(\sigma) \qquad X = Y \Rightarrow Y = X .$$

Lemma 2.20. Weak equality is the equivalence relation generated by weak reducibility. (Cf. Lemma 1.12.)

The following theorem and its corollaries correspond to Theorem 1.13 etc., and have the same importance.

Theorem 2.21. If $X = Y$, then there is a Z such that $X \triangleright Z$ and $Y \triangleright Z$.

Corollary 2.21.1. If $X = Y$ and Y is in normal form, then $X \triangleright Y$.

Corollary 2.21.2. If $X = Y$, then either X and Y do not have normal forms, or they both have the same normal form.

Corollary 2.21.3. If X and Y are both in normal form, then $X \not\equiv Y \Rightarrow X \neq Y$.

This concludes the basic properties of combinators.

Remark 2.22. The reduction and equality defined here do not correspond exactly to those of λ-conversion. For instance, λ-equality has the property

$$X = Y \Rightarrow \lambda x. X = \lambda x. Y$$

(by Definition 1.11(ξ)), but the corresponding statement for combinators,

$$(\xi) \qquad X = Y \Rightarrow [x]X = [x]Y ,$$

is not true in general. For example

$$\mathbf{S}xyz = xz(yz) ,$$

but $[x]\mathbf{S}xyz \neq [x]xz(yz)$; in fact we have

$$[x]\mathbf{S}xyz \equiv \mathbf{S}(\mathbf{SS}(\mathbf{K}y))(\mathbf{K}z) ,$$
$$[x]xz(yz) \equiv \mathbf{S}(\mathbf{SI}(\mathbf{K}z))(\mathbf{K}(yz)) ,$$

which are unequal by Corollary 2. 21. 3.

This same example shows that equality is not extensional in the sense of Remark 1. 14, since $([x]\mathbf{S}xyz)V = ([x]xz(yz))V$ for all V. Chapters 5 to 7 will look at extensional equality and its corresponding reduction. These are needed for combinator-based logic (Chapter 10), but not for representing recursive functions, etc. (Chapters 3, 4, 11.)

Remark 2. 23 (Interpreting the terms). Just as with λ-terms, the exact interpretation varies with different applications of the theory. In general the terms are intended to represent one-place functions, and XY represents the result of applying X to the argument Y. But some of the terms may be left uninterpreted (e. g. perhaps XX, depending on the application in question), and in some contexts it is only the formal properties of reduction and equality that are needed, without any interpretation.

The combinators **I**, **K**, **S** may be interpreted as at the start of this chapter, though for example, **K** does not represent an operation on a particular set of objects (in the usual sense), but rather the abstract notion of forming a constant-function from any object. This is not easy to interpret set-theoretically; most set-theories implicitly embody some sort of type-restrictions, so operations which apply to all objects are usually excluded. But one of Curry's early ideas behind combinatory logic was that it might be possible to use this more abstract concept of operation as a basis for mathematics instead of the set concept. For example it seems a bit unnatural to confine the very simple notions of identity-operation and constant-forming operation (represented by **I** and **K**) to a particular domain; perhaps our uneasiness with such less confined notions is merely due to our being used to thinking in terms of set theory. From this point of view the problem of interpreting combinators as sets would become less important, and be replaced by the problem of actually formulating a system of logic based on a less confined notion of operation. (Of course to avoid contradictions there would have to be some restrictions, just as formal set theories need to restrict the crude notion of set.) This problem will be considered in Chapter 10.

However, despite the above comments, it is obviously interesting to see just how far combinators can be interpreted in the usual kind of set theory. The simplest way (cf. Chapters 8 and 9) is to introduce type-restrictions and only to interpret terms satisfying these restrictions. A more ingenious interpretation, in which terms are interpreted as sequences of functions and no term is left uninterpreted, has recently been devised by D. Scott in [LTM], in connection with his study of the semantics of computer languages.

Also, interpretations in a modified sense are provided by the 'uniformly reflexive structures' of Strong [AGR] and Wagner [URS], and related systems of recursion theory.

Remark 2.24 (History). Combinators were first introduced by M. Schönfinkel in 1920 (see his [BML]). They were rediscovered independently a few years later by Curry, who has been responsible for the main development of combinatory logic since then (although others, such as Rosser, have also been involved at various times). The development of λ-conversion was started at about the same time, by Church, Rosser and Kleene.

At present (1971), combinators have three main uses: first, for constructing logical systems based on the abstract notion of operation suggested in the previous remark; second, as the basis of a notation for constructive functions of various kinds in proof-theory; and third, in the construction and analysis of some programming languages in computer science.

The first kind of work will be introduced in Chapters 9 and 10 of these notes, and further development is in [CLg. II]. A similar type of work has also been carried out by Fitch (see his [SCD] and the references therein).

An example of the second application will be found in Chapter 11, and other examples are in the references quoted there. A proof-theoretic example of a slightly different kind is in Goodman [IAT].

As mentioned in the introduction, the third application will not be covered here. However, some of the relevant papers on the topic are Böhm and Gross [ICC], Landin [MEE], [CAC], Strachey [TFS], [FCP], Scott [LTM], [OMT], and Orgass and Fitch [TCM], [SRP].

Finally, it is worth mentioning that λ-terms have proved a natural notation for formal theories of types; see Church [FST], Henkin [CTT], Andrews [TTT].

Exercises

1. What is the difference between

$[M_1/x_1, \ldots, M_n/x_n]Y, \quad [M_1/x_1][M_2/x_2]\ldots[M_n/x_n]Y$?

2. In terms of **I**, **K**, **S**, construct terms **W**, **Φ**, such that

$$\mathbf{W}XY \triangleright XYY\,,$$
$$\mathbf{\Phi}XYZU \triangleright X(YU)(ZU)\,.$$

3. Prove that a term X is in normal form if and only if X is irreducible; i.e. iff

$$X \triangleright Y \Rightarrow Y \equiv X\,.$$

Show that this would be false if there were an atom **W** with an axiom-scheme $\mathbf{W}XY \triangleright XYY$.

4. Show that $X \triangleright Y \triangleright X$ does not imply $Y \equiv X$.

5. Construct a term $\mathbf{Y}$ such that $\mathbf{YX} = X(\mathbf{YX})$ for all X. Such $\mathbf{Y}$ are called fixed-point combinators, since they operate on X to produce a fixed-point of X. (Hint: try $\mathbf{Y} \equiv VV$ with a suitable V; see [CLg. I] §5G, [CLg. II] §11F7 for answers.)

6. Show that adding $\mathbf{S} = \mathbf{K}$ as a new axiom to Definition 2.19 will produce an equality in which all terms are equal. (This result can be proved from first principles. But if equality is extensional (see Chapter 5), it is easier to deduce it from the following interesting result in C. Böhm's [PFB]: To each pair X, Y of unequal combinators with normal forms there corresponds a combinator D such that

$$DUVX =_e U,$$
$$DUVY =_e V$$

for all U and V. (Here '$=_e$' denotes extensional equality.))

7. Show that there is no combinator D which distinguishes between atoms and composite terms; i.e. there is no D such that

$$DX = \mathbf{S} \quad \text{if } X \text{ is an atom,}$$
$$DX = \mathbf{K} \quad \text{if } X \text{ has form } UV.$$

3. Representing the Recursive Functions

If we pick out a sequence of combinators to represent the natural numbers, then some of the other combinators will represent certain functions of natural numbers. In this chapter we shall make this representation precise, and see that the functions so representable are just the partial recursive functions.

Notation. The terms here will be combinatory terms, not λ-terms. Letters 'i', 'j', 'k', 'm' and 'n' will denote non-negative natural numbers. An n-place function ϕ of natural numbers will be called <u>properly partial</u> iff $\phi(m_1, \ldots, m_n)$ is undefined for some $m_1, \ldots, m_n$, and <u>total</u> otherwise. As usual, a <u>partial</u> function may be either properly partial or total. For any terms X, Y, we shall use the abbreviation

$$X^n Y \equiv \underbrace{X(X(\ldots(XY)\ldots))}_{n\ Xs} \qquad \text{for } n > 0,$$

$$X^0 Y \equiv Y.$$

Definition 3.1. We shall represent each natural number n by the term

$$\mathbf{Z}_n \equiv (\mathbf{SB})^n(\mathbf{KI}).$$

These terms will be called <u>numerals</u>. There are many other sequences that would do instead of $\mathbf{Z}_n$, but this one has the technical advantage that

$$(3.2) \qquad \mathbf{Z}_n FX \rhd F^n X$$

for all F and X. (Compare Church's numerals, [CLC] p. 28.)

Definition 3.3. Let ϕ be an n-argument function of natural numbers. Then a term X combinatorially defines ϕ in the weak sense iff for all $m_1, \ldots, m_n$,

$$XZ_{m_1} \ldots Z_{m_n} = Z_{\phi(m_1, \ldots, m_n)}$$

whenever $\phi(m_1, \ldots, m_n)$ exists. If, in addition, the left-hand term has no normal form when $\phi(m_1, \ldots, m_n)$ does not exist, we shall say that X defines ϕ in the strong sense.

Theorem 3.8 will show that every partial recursive function can be defined combinatorially in the weak sense. They can also be defined strongly, but to avoid the technicalities of existence of normal forms we shall only prove the weaker result here. (See [CLg. II] §13A4, especially the remarks after Theorem 5.) Of course for total functions the two results are the same. We shall also leave out the proof that every strongly definable function is partial recursive, since the proof proceeds by Gödel-numbering techniques which are tedious but fairly standard. (See Kleene [LDR].)

Theorem 3.4 (Essentially due to Kleene). Every primitive recursive function ϕ can be combinatorially defined by a term $\overline{\phi}$.

Proof. For notational convenience we shall include the natural numbers as 0-place functions. Then in [IMM] §44 Remark 1 (Basis B), Kleene has shown that the class of all primitive recursive functions can be defined inductively as follows:

(I) The successor function σ is primitive recursive.

(II) 0 is a 0-place primitive recursive function.

(III) If

$$\phi(m_1, \ldots, m_n) \equiv m_k$$

for all $m_1, \ldots, m_n$ and some $k \leq n$, then ϕ is primitive recursive.

(IV)   If

$$\phi(m_1, \ldots, m_n) \equiv \psi(\chi_1(m_1, \ldots, m_n), \ldots, \chi_p(m_1, \ldots, m_n))$$

and ψ, χ_1, $\ldots$, χ_p are primitive recursive, then ϕ is primitive recursive.

(V)   If

$$\phi(0, m_1, \ldots, m_n) \equiv \psi(m_1, \ldots, m_n) \quad \text{and}$$

$$\phi(k+1, m_1, \ldots, m_n) \equiv \chi(k, \phi(k, m_1, \ldots, m_n), m_1, \ldots, m_n)$$

and ψ and χ are primitive recursive, then ϕ is primitive recursive.

The term $\overline{\phi}$ is chosen by a corresponding induction, as follows:

(I)   Let $\overline{\sigma} \equiv \mathbf{SB}$.

(II)   Let $\overline{0} \equiv \mathbf{Z}_0$.

(III)   Let $\overline{\phi} \equiv [x_1, \ldots, x_n]x_k$.

(IV)   Given that $\overline{\psi}$, $\overline{\chi}_1$, $\ldots$, $\overline{\chi}_p$ define ψ, χ_1, $\ldots$, χ_p respectively, let

$$\overline{\phi} \equiv [x_1, \ldots, x_n](\overline{\psi}(\overline{\chi}_1 x_1 \cdots x_n) \cdots (\overline{\chi}_p x_1 \cdots x_n)).$$

(V)   Suppose we could construct a combinator $\mathbf{R}$ such that for all X, Y, k,

(3.5) $$\begin{cases} \mathbf{R}XYZ_0 = X, \\ \mathbf{R}XYZ_{k+1} = YZ_k(\mathbf{R}XYZ_k). \end{cases}$$

Then, given that $\overline{\psi}$ and $\overline{\chi}$ define ψ and χ respectively, we could choose

$$\phi \equiv [u, x_1, \ldots, x_n]\mathbf{R}(\overline{\psi}x_1 \cdots x_n)(Y^*x_1 \cdots x_n)u ,$$

where

$$Y^* \equiv [x_1, \ldots, x_n, u, v]\overline{\chi}uvx_1 \cdots x_n .$$

This $\overline{\phi}$ would define ϕ, because for all X_1, $\ldots$, X_n,

$$\bar{\phi}\mathbf{Z}_0 X_1 \ldots X_n = \mathbf{R}(\bar{\psi}X_1 \ldots X_n)(Y^*X_1 \ldots X_n)\mathbf{Z}_0 \quad \text{by Thm. 2.16}$$
$$= \bar{\psi}X_1 \ldots X_n \quad \text{by (3.5)}$$

$$\bar{\phi}\mathbf{Z}_{k+1}X_1 \ldots X_n = \mathbf{R}(\bar{\psi}X_1 \ldots X_n)(Y^*X_1 \ldots X_n)\mathbf{Z}_{k+1}$$
$$= Y^*X_1 \ldots X_n\mathbf{Z}_k(\mathbf{R}(\bar{\psi}X_1 \ldots X_n)(Y^*X_1 \ldots X_n)\mathbf{Z}_k)$$
$$\text{by (3.5)}$$
$$= Y^*X_1 \ldots X_n\mathbf{Z}_k(\bar{\phi}\mathbf{Z}_k X_1 \ldots X_n)$$
$$= \bar{\chi}\mathbf{Z}_k(\bar{\phi}\mathbf{Z}_k X_1 \ldots X_n)X_1 \ldots X_n \quad \text{by def. of } Y^*.$$

Thus to complete the proof we only need to construct an **R** satisfying (3.5). To see how such an **R** can be built up, consider for example a primitive recursive function ϕ defined by

$$\phi(0) \equiv m, \qquad \phi(k+1) \equiv \chi(k, \phi(k)) .$$

To calculate $\phi(k)$ we may first write down the ordered pair $(0, m)$ and then iterate k times the operation f such that

$$f((n, \phi(n))) \equiv (n+1, \chi(n, \phi(n))) ,$$

and finally take the second member of the last ordered pair produced. The **R** defined below will just be the combinatory analogue of this calculation process.

First define

$$(3.6) \qquad \mathbf{D} \equiv [x, y, z]z(\mathbf{K}y)x .$$

Then for all X, Y, k we have by Definition 3.1 and Theorem 2.16,

$$(3.7) \qquad \begin{cases} \mathbf{D}XY\mathbf{Z}_0 = X, \\ \mathbf{D}XY\mathbf{Z}_{k+1} = Y . \end{cases}$$

Thus we can take $\mathbf{D}\mathbf{Z}_m\mathbf{Z}_n$ as the combinatory analogue of the ordered pair (m, n), since (3.7) gives us a method of picking out the first or second element. Now define

$$Q \equiv [y, v]\mathbf{D}(\mathbf{SB}(v\mathbf{Z}_0))(y(v\mathbf{Z}_0)(v\mathbf{Z}_1)) .$$

Then for any Y, X, n,

$$\begin{aligned} QY(\mathbf{DZ}_n X) &= \mathbf{D}(\mathbf{SB}(\mathbf{DZ}_n X\mathbf{Z}_0))(Y(\mathbf{DZ}_n X\mathbf{Z}_0)(\mathbf{DZ}_n X\mathbf{Z}_1)) \\ &= \mathbf{D}(\mathbf{SBZ}_n)(Y\mathbf{Z}_n X) \qquad \text{by (3.7)} \\ &= \mathbf{DZ}_{n+1}(Y\mathbf{Z}_n X)\ . \end{aligned}$$

Thus QY imitates the operation f above, if Y is a term defining χ. Also, for all X, Y and k we have

$$(QY)^k(\mathbf{DZ}_0 X) = \mathbf{DZ}_k X_k$$

for some term X_k (which in fact corresponds to the value of $\phi(k)$ above, if Y defines χ and $X \equiv \mathbf{Z}_m$).

Now define

$$\mathbf{R} \equiv [x,\ y,\ u]u(Qy)(\mathbf{DZ}_0 x)\mathbf{Z}_1\ .$$

Then

$$\begin{aligned} \mathbf{R}XY\mathbf{Z}_k &= \mathbf{Z}_k(QY)(\mathbf{DZ}_0 X)\mathbf{Z}_1 \\ &= (QY)^k(\mathbf{DZ}_0 X)\mathbf{Z}_1 \quad \text{by (3.2)} \\ &= \mathbf{DZ}_k X_k \mathbf{Z}_1 \quad \text{as above} \\ &= X_k \quad \text{by (3.6).} \end{aligned}$$

Hence the above **R** does satisfy (3.5), because

$$\begin{aligned} \mathbf{R}XY\mathbf{Z}_0 &= (QY)^0(\mathbf{DZ}_0 X)\mathbf{Z}_1 \\ &= \mathbf{DZ}_0 X\mathbf{Z}_1 \quad \text{by definition of } (QY)^0 \\ &= X \quad \text{By (3.7);} \end{aligned}$$

$$\begin{aligned} \mathbf{R}XY\mathbf{Z}_{k+1} &= (QY)^{k+1}(\mathbf{DZ}_0 X)\mathbf{Z}_1 \\ &= (QY)((QY)^k(\mathbf{DZ}_0 X))\mathbf{Z}_1 \quad \text{by definition of } (QY)^{k+1} \\ &= QY(\mathbf{DZ}_k X_k)\mathbf{Z}_1 \quad \text{as above} \\ &= \mathbf{DZ}_{k+1}(Y\mathbf{Z}_k X_k)\mathbf{Z}_1 \\ &= Y\mathbf{Z}_k X_k \\ &= Y\mathbf{Z}_k(\mathbf{R}XY\mathbf{Z}_k)\ . \end{aligned}$$

Thus the theorem is proved.

Theorem 3.8 (Kleene). Every partial recursive function ϕ can be combinatorially defined by a term $\overline{\phi}$, in the weak sense.

Proof. By Kleene [IMM] §58, there exist primitive recursive functions ψ and χ such that

$$\phi(m_1, \ldots, m_n) \equiv \psi(\mu k[\chi(m_1, \ldots, m_n, k) \equiv 0]),$$

where for each $m_1, \ldots, m_n$, $\mu k[\chi(m_1, \ldots, m_n, k) \equiv 0]$ is the least k, if any, for which $\chi(m_1, \ldots, m_n, k) \equiv 0$, and is undefined if no such k exists.

Another way of looking at μ is to define, for each k,

$$\theta(m_1, \ldots, m_n, k) \equiv k \text{ if } \chi(m_1, \ldots, m_n, k) \equiv 0,$$
$$\equiv \theta(m_1, \ldots, m_n, k+1) \text{ otherwise.}$$

Then

$$\theta(m_1, \ldots, m_n, 0) \equiv \mu k[\chi(m_1, \ldots, m_n, k) \equiv 0].$$

Hence,

$$\phi(m_1, \ldots, m_n) \equiv \psi(\theta(m_1, \ldots, m_n, 0)).$$

Now by Theorem 3.4, the primitive recursive functions ψ and χ can be combinatorially defined by certain terms $\overline{\psi}$ and $\overline{\chi}$, so to define ϕ combinatorially it is enough to so define θ.

Begin by defining

$$P \equiv [x, y]Tx(xy)(Tx)y$$

where $T \equiv [x]\mathbf{D}(\mathbf{KI})([u]\mathbf{B}([v]u(xv)uv)(\mathbf{SB}))$. Then we have

$$(3.9) \qquad \begin{cases} PXY = Y \text{ if } (XY) = \mathbf{Z}_0, \\ PXY = PX(\mathbf{SB}Y) \text{ if } (XY) = \mathbf{Z}_{m+1} \text{ for some } m. \end{cases}$$

Proof of (3.9).

$$PXY = TX(XY)(TX)Y$$
$$= \mathbf{D}(\mathbf{KI})([u]\mathbf{B}([v]u(Xv)uv)(\mathbf{SB}))(XY)(TX)Y,$$

for variables u, v not occurring in X. (X can be substituted for x in [v]u(xv)uv by Lemma 2.17(b).) Therefore, if $(XY) = \mathbf{Z}_0$,

$$\begin{aligned} PXY &= \mathbf{KI}(TX)Y \quad \text{by } (3.7) \\ &= Y , \end{aligned}$$

and if $(XY) = \mathbf{Z}_{m+1}$,

$$\begin{aligned} PXY &= ([u]\mathbf{B}([v]u(Xv)uv)(\mathbf{SB}))(TX)Y \quad \text{by } (3.7) \\ &= \mathbf{B}([v]TX(Xv)(TX)v)(\mathbf{SB})Y \quad \text{by Theorem 2.12, Lemma 2.17(b)} \\ &= ([v]TX(Xv)(TX)v)(\mathbf{SB}Y) \quad \text{by Example 2.5} \\ &= TX(X(\mathbf{SB}Y))(TX)(\mathbf{SB}Y) \quad \text{by Theorem 2.12} \\ &= PX(\mathbf{SB}Y) . \end{aligned}$$

This completes the proof of (3.9).

Now, returning to the proof of the main theorem, define

$$\bar{\theta} \equiv [x_1, \ldots, x_n]P(\bar{\chi}x_1 \ldots x_n) .$$

Then

$$\bar{\theta}\mathbf{Z}_{m_1} \ldots \mathbf{Z}_{m_n}\mathbf{Z}_k = P(\bar{\chi}\mathbf{Z}_{m_1} \ldots \mathbf{Z}_{m_n})\mathbf{Z}_k .$$

If the term in parentheses reduces to $\mathbf{Z}_0$, then the right-hand side of this equation is equal to $\mathbf{Z}_k$ by (3.9). On the other hand if the term in parentheses reduces to $\mathbf{Z}_{m+1}$, (3.9) shows that

$$\begin{aligned} \bar{\theta}\mathbf{Z}_{m_1} \ldots \mathbf{Z}_{m_n}\mathbf{Z}_k &= P(\bar{\chi}\mathbf{Z}_{m_1} \ldots \mathbf{Z}_{m_n})\mathbf{Z}_{k+1} \\ &= \bar{\theta}\mathbf{Z}_{m_1} \ldots \mathbf{Z}_{m_n}\mathbf{Z}_{k+1} . \end{aligned}$$

Thus $\bar{\theta}$ combinatorially defines θ, completing the proof.

Remark 3.10. Instead of representing the numbers by combinators $\mathbf{Z}_n$, for some purposes it is more convenient to add two new constant atoms $\bar{0}$, $\bar{\sigma}$ to the definition of the terms, and represent each number n by

$$\overline{n} \equiv \overline{\sigma}^{n}\overline{0} .$$

In this case an **R** with property (3. 5) cannot be constructed (see [CLg. II], §13A3 Theorem 2). But **R** can be added as a third new atom, with new axiom-schemes

$$(3.11) \qquad \begin{cases} \mathbf{RXY}\overline{0} \triangleright \mathbf{X} , \\ \mathbf{RXY}(\overline{k+1}) \triangleright \mathbf{Y}\overline{k}(\mathbf{RXY}\overline{k}) \end{cases}$$

added to the definitions of reduction and equality. Then the proofs in this chapter can be carried out for the new system. For example, a term $\mathbf{D}^*$ satisfying (3. 7) with $\overline{0}$, $\overline{k+1}$ instead of $\mathbf{Z}_0$, $\mathbf{Z}_{k+1}$ can be defined by

$$\mathbf{D}^* \equiv [x, y]\mathbf{R}x(\mathbf{K}(\mathbf{K}y)) ,$$

and a P satisfying (3. 9) with $\overline{0}$, $\overline{m+1}$ instead of $\mathbf{Z}_0$, $\mathbf{Z}_{m+1}$ can be defined by replacing **SB** by $\overline{\sigma}$ in the previous definition of P, and using $\mathbf{D}^*$ instead of **D**.

Instead of adding **R** as an atom, we could add an atom **Z** with the simpler axiom-scheme of reduction

$$(3.12) \qquad \mathbf{Z}\overline{n} \triangleright \mathbf{Z}_n .$$

Then **R** could be defined in terms of **Z** in several ways. For example, using the abbreviation

$$Q \equiv [y, v]\mathbf{D}(\overline{\sigma}(v\mathbf{Z}_0))(y(v\mathbf{Z}_0)(v\mathbf{Z}_1)) ,$$

we could define

$$R \equiv [x, y, u]\mathbf{Z}u(Qy)(\mathbf{D}\overline{0}x)\mathbf{Z}_1 .$$

This is nearly the same **R** as in the proof of Theorem 3. 4, and almost the same proof as before shows that it satisfies (3. 11). Hence the partial recursive functions can be defined in the system with (3. 12) instead of (3. 11).

Incidentally, reduction modified by adding (3. 11) or (3. 12) still has all the usual properties from Chapter 2, including

Theorem 2.18 and its consequences; see the end of Appendix 1, and [CLg. II] §12A3-4.

Remark 3.13. The proofs in this chapter apply to λ-terms if we replace **S**, **K**, **I** respectively by

$$\mathbf{S}_\lambda \equiv \lambda xyz.\, xz(yz), \quad \mathbf{K}_\lambda \equiv \lambda xy.\, x, \quad \mathbf{I}_\lambda \equiv \lambda x.\, x,$$

and replace '$[x_1, \ldots, x_n]$' by '$\lambda x_1 \ldots x_n$', etc. Thus all the partial recursive functions can be defined as λ-terms.

Exercises

1. Construct terms which define the following functions:

(a) Addition.

(b) Multiplication.

(c) The function $\dot{-}$, defined by setting $m \dot{-} n \equiv m - n$ if $m \geq n$, but $m \dot{-} n \equiv 0$ if $m < n$.

(d) The function e, defined by setting $e(m, n) \equiv 0$ if $m \equiv n$, but $e(m, n) \equiv 1$ if $m \not\equiv n$.

2. Show that if the numerals are defined in terms of atoms $\bar{0}$ and $\bar{\sigma}$ as in Remark 3.10, but **R** is not present, we cannot even define the predecessor function; that is, there is no term P such that

$$P\bar{0} = \bar{0}, \qquad P(\overline{n+1}) = \bar{n}.$$

(Hint: Substitute suitable terms for $\bar{0}$ and $\bar{\sigma}$ so that the equations for P violate Corollary 2.21.3. See [CLg. II] §13A3 Theorem 2 for a solution.)

3. If a function ϕ is combinatorially defined in the weak sense, is ϕ necessarily partial recursive?

4. The Undecidability Theorem

The aim of this chapter is to prove a general undecidability theorem which shows in particular that the '=' relation is recursively undecidable, and that there is no recursive way of deciding whether a term has a normal form or not. The analogue of these results for Church's λ-conversion gave Church his proof of the undecidability of pure first-order predicate calculus. The present proof will be carried out for combinators, but with appropriate changes it will also be valid for λ-terms. (Cf. Remark 3.13.) It will not be valid for the restricted λ-terms of Church's original system (Remark 1.15).

The theorem will be proved by the device of Gödel numbering. As in Definition 3.1, the numerals are defined by

$$\overline{n} \equiv \mathbf{Z}_n \equiv (\mathbf{SB})^n(\mathbf{KI}) .$$

We can assign to each term X a number n, called gd(X) or the Gödel number of X. The corresponding numeral $\overline{n}$ will be called $\ulcorner X \urcorner$. The numbering can be done in many different ways, for example as in Kleene [IMM] §52, but we need not specify exactly which numbering is used, since all we shall need to know about it are the following two properties:

(4.1) There is a recursive function μ such that for all terms X and Y,

$$\mu(\text{gd}(X), \text{gd}(Y)) \equiv \text{gd}(XY) .$$

(4.2) There is a recursive function ν such that for all natural numbers n,

$$\nu(n) \equiv \text{gd}(\overline{n}) .$$

The proof that there is a numbering with properties (4.1) and (4.2) will be omitted, but the essential idea is that the operation of forming (XY) from X and Y is effectively computable, and so is the operating of forming $\bar{n}$ from n.

Definition 4.3. Two sets $\mathscr{A}$ and $\mathscr{B}$ of natural numbers are called <u>recursively separable</u> iff there is a total recursive function ϕ whose only values are 0 and 1, such that

$$n \in \mathscr{A} \Rightarrow \phi(n) \equiv 1 ,$$
$$n \in \mathscr{B} \Rightarrow \phi(n) \equiv 0 .$$

Two sets $\mathscr{A}$ and $\mathscr{B}$ of terms are <u>recursively separable</u> iff the corresponding sets of Gödel numbers are recursively separable.

Definition 4.4. A set $\mathscr{A}$ of terms is <u>closed under equality</u> iff, for all terms X and Y,

$$X \in \mathscr{A} \text{ and } Y = X \Rightarrow Y \in \mathscr{A} .$$

<u>Theorem 4.5.</u> <u>No pair of non-empty sets of terms which are closed under equality is recursively separable.</u>

Proof. Suppose that ϕ separates $\mathscr{A}$ and $\mathscr{B}$, where $\mathscr{A}$ and $\mathscr{B}$ are disjoint sets of terms both of which are non-empty and closed under equality, and let F be a term which defines ϕ, then we have

$$X \in \mathscr{A} \Rightarrow F\ulcorner X\urcorner = \bar{1} ,$$
$$X \in \mathscr{B} \Rightarrow F\ulcorner X\urcorner = \bar{0} .$$

Let M and N be terms which define the functions μ and ν of (4.1) and (4.2) respectively; then for all X, Y, and n,

$$M\ulcorner X\urcorner\ulcorner Y\urcorner = \ulcorner XY\urcorner ,$$
$$N\bar{n} = \ulcorner\bar{n}\urcorner .$$

Choose any term U in $\mathscr{A}$ and any term V in $\mathscr{B}$. We shall construct a term J such that

$$F\ulcorner J\urcorner = \bar{1} \;\Rightarrow\; J = V\,, \qquad \text{and}$$
$$F\ulcorner J\urcorner = \bar{0} \;\Rightarrow\; J = U\,.$$

This will cause a contradiction because, letting $j \equiv gd(J)$, we have

$$\begin{aligned} \phi(j) \equiv 1 \;&\Rightarrow\; F\ulcorner J\urcorner = \bar{1} \\ &\Rightarrow\; J = V \\ &\Rightarrow\; J \in \mathscr{B} \qquad \text{since } \mathscr{B} \text{ is closed under equality} \\ &\Rightarrow\; \phi(j) \equiv 0\,; \end{aligned}$$

and

$$\begin{aligned} \phi(j) \equiv 0 \;&\Rightarrow\; F\ulcorner J\urcorner = \bar{0} \\ &\Rightarrow\; J = U \\ &\Rightarrow\; J \in \mathscr{A} \quad \text{since } \mathscr{A} \text{ is closed under equality} \\ &\Rightarrow\; \phi(j) \equiv 1\,, \end{aligned}$$

and ϕ is a total function whose only values are 0 and 1.

It will be enough to construct J such that

$$J = \mathbf{D}UV(F\ulcorner J\urcorner)$$

where $\mathbf{D}$ is defined by (3. 6), since this $\mathbf{D}$ has the property that

$$\mathbf{D}UV\bar{1} = V\,,$$
$$\mathbf{D}UV\bar{0} = U\,.$$

To do this, define

$$J \equiv H\ulcorner H\urcorner\,,$$

where $H \equiv [y]\mathbf{D}UV(F(My(Ny)))$.

Then

$$\begin{aligned} J &= \mathbf{D}UV(F(M\ulcorner H\urcorner(N\ulcorner H\urcorner))) && \text{by definition of } H, \\ &= \mathbf{D}UV(F(M\ulcorner H\urcorner\ulcorner\ulcorner H\urcorner\urcorner)) && \text{by definition of } N, \\ &= \mathbf{D}UV(F\ulcorner(H\ulcorner H\urcorner)\urcorner) && \text{by definition of } M, \\ &= \mathbf{D}UV(F\ulcorner J\urcorner) && \text{by definition of } J, \end{aligned}$$

as required.

This proof is adapted from Curry [CLg. II] §13B2 and unpublished notes by D. Scott (Stanford, 1963).

Corollary 4.5.1. A non-empty set $\mathscr{A}$ of terms which is closed under equality cannot be recursive.

Proof. In the theorem, let $\mathscr{B}$ be the complement of $\mathscr{A}$.

Corollary 4.5.2. The set of all terms which have normal forms is not recursive.

Corollary 4.5.3. The relation of equality between terms is not recursive: that is, there is no total recursive function ψ such that

$$\psi(gd(X),\ gd(Y)) \equiv \begin{cases} 1 & \text{if } X = Y \\ 0 & \text{if } X \neq Y . \end{cases}$$

Proof. In Corollary 4.5.1, let $\mathscr{A}$ be the set of all terms equal to one particular term (**I**, for example).

Remark. Church's proof of undecidability of pure classical first-order predicate calculus can be summarized as follows. When the combinatory terms are Gödel-numbered, equality of terms corresponds to a relation between the natural numbers. Numbers can be represented in a pure predicate calculus which has function-symbols by choosing a variable z and a function-symbol f and letting

z represent 0,
f(z) represent 1,
f(f(z)) represent 2, etc.

The eight axiom-schemes and rules defining equality of terms can by this means be translated into predicate-calculus formulae $\mathscr{F}_1, \ldots, \mathscr{F}_8$ containing a predicate-letter **E**, such that for all predicate-calculus-terms s and t representing numbers n_s and n_t respectively, $(\mathcal{F}_1 \& \ldots \& \mathcal{F}_8) \supset E(s, t)$ is provable iff n_s and n_t are Gödel numbers of equal combinatory terms. Hence if we could decide all questions of provability in the pure predicate calculus, we could then decide whether arbitrary combinatory terms are equal, contrary to Corollary 4. 5. 3. (The details of this proof are in Church [NEP], [UPE].)

For some other undecidability theorems about reduction and equality, see Barendregt [SEM] pp. 24-26.

5. Extensional Equality

A relation = is said to be extensional if the following rule (ext) is valid for it:

(ext) If $XV = YV$ for all V, then $X = Y$.

We have observed (Remarks 1. 14 and 2. 22) that the equality relations of Chapters 1 and 2 are not extensional. In this chapter we consider several alternative formulations of extensional equality and prove their equivalence. (Note that (ext) is an infinite rule in the sense that it involves infinitely many premises. The alternative formulations avoid infinite rules.)

Remark 5. 1. Note that in rule (ext) the terms V may contain (free) occurrences of variables. If one intuitively thinks of the variables as ranging over the domain of terms with fixed meanings - i. e. terms without (free) occurrences of variables - then it is plausible to also restrict the terms V in (ext) to that domain. The result of this restriction would be the following rule:

(ext') If $XV = YV$ for all V containing no free variables, then $X = Y$.

It is not known, however, whether rule (ext') is equivalent to rule (ext). It may even be that there are several inequivalent rules (ext'), depending on the system's set of constant atoms. However, we shall not consider (ext') here.

Notation. We will use the equality symbol '=' in giving the several formulations of extensionality, and we shall indicate specific such relations by appropriate subscripts (Definition 5. 2 and Definition 5. 6). Recall that λ-equality was defined inductively in Chapter 1 by the following axiom-schemes and rules (Definition 1. 11):

$$(\alpha) \quad \lambda y.\, z = \lambda v.\, [v/y]z, \quad (y \notin Z;\ v, y \text{ not bound in } Z)$$
$$(\beta) \quad (\lambda x.\, M)N = [N/x]M\ ,$$
$$(\rho) \quad X = X\ ,$$
$$(\mu), (\nu) \quad X = X' \Rightarrow ZX = ZX' \text{ and } XZ = X'Z\ ,$$
$$(\tau) \quad X = Y \text{ and } Y = Z \Rightarrow X = Z\ ,$$
$$(\sigma) \quad X = Y \Rightarrow Y = X\ ,$$
$$(\xi) \quad X = X' \Rightarrow \lambda x.\, X = \lambda x.\, X'\ .$$

Combinatory equality was defined inductively by the axiom-schemes

$$\mathbf{S}XYZ = XZ(YZ)\ ,$$
$$\mathbf{K}XY = X\ ,$$
$$\mathbf{I}X = X\ ,$$
$$X = X\ ,$$

together with rules (μ), (ν), (τ) and (σ). (See Definition 2. 19.) Notice that 'defined inductively' means that to each true statement $X = Y$ corresponds a tree-form deduction of $X = Y$ from the clauses of the definition. The total number of applications of clauses used in this tree will be called the length of the deduction.

Now consider the following: rule (ζ), for either λ-terms or combinatory terms; rule (ξ), for combinatory terms; and axiom scheme (η), for λ-terms.

(ζ) If $Xx = Yx$ for any variable $x \notin XY$, then $X = Y$.

(ξ) If $X = Y$, then for any variable x, $[x]X = [x]Y$.

(η) $\lambda x.\, Xx = X$ if $x \notin X$.

Definition 5. 2 (Alternative formulations of extensional equality). The relation obtained by adding (ext) to either Definition 1. 11 or Definition 2. 19 is denoted by $=_e$. The relation $=_\zeta$ is defined by adding (ζ) to Definition 1. 11 or Definition 2. 19. The relation $=_\xi$ is defined by Definition 2. 19

together with (ξ), and $=_\eta$ is defined by Definition 1.11 together with (η).

Lemma 5.3. Suppose $X =_\zeta Y$ by a proof of length n. Let m be any integer > 0 and let X', Y' be the results of the simultaneous substitution of terms $Z_1, \ldots, Z_m$ for distinct variables $x_1, \ldots, x_m$ in X, Y respectively. Then $X' =_\zeta Y'$ by a proof of length n.

Proof. By induction on n.

Notice that this lemma applies to both combinatory and λ-terms. Simultaneous substitution is defined for the former in Definition 2.2, and for the latter it is a natural modification of Definition 1.4.

Theorem 5.4. $X =_e Y$ iff $X =_\zeta Y$. Moreover, if X, Y are combinatory terms, $X =_\zeta Y$ iff $X =_\xi Y$. If X, Y are λ-terms, $X =_\zeta Y$ iff $X =_\eta Y$.

Proof. To prove the theorem for combinatory terms, we show

$$X =_e Y \Rightarrow X =_\xi Y \Rightarrow X =_\zeta Y \Rightarrow X =_e Y .$$

In order to show that $X =_e Y$ implies $X =_\xi Y$, it suffices to show that (ext) is a valid rule for $=_\xi$. To do this, suppose that $XV =_\xi YV$ for all V. Among the terms V is a variable $x \notin XY$. Thus, $Xx =_\xi Yx$ for $x \notin XY$. By (ξ), we conclude

$$[x]Xx =_\xi [x]Yx ;$$

that is, $X =_\xi Y$ as required.

To show $X =_\xi Y$ implies $X =_\zeta Y$, we must show that

$$X =_\zeta Y \Rightarrow [x]X =_\zeta [x]Y .$$

Note that $([x]X)x = X$ without postulating any principle of extensionality, so

$$X =_\zeta Y \Rightarrow ([x]X)x =_\zeta ([x]Y)x .$$

Since $x \notin ([x]X)([x]Y)$, by (ζ) we conclude $[x]X =_\zeta [x]Y$, as desired.

To show $X =_\zeta Y$ implies $X =_e Y$, we use induction on the length n of the proof of $X =_\zeta Y$. The cases when $X =_\zeta Y$ is either an axiom or a rule other than (ζ) are trivial, since the axioms and these rules are part of the definition of $=_e$. Suppose then that $X =_\zeta Y$ is a consequence of (ζ). Hence there is a variable $z \notin XY$ such that $Xz =_\zeta Yz$ by a proof of length $n - 1$. By Lemma 5. 3, for any V there is a proof of length $n - 1$ of $XV =_\zeta YV$. Therefore by hypothesis of induction,

$$XV =_e YV$$

for all V, and so we conclude $X =_e Y$. This completes the proof of the theorem for combinatory terms.

To prove the theorem for λ-terms, we show that

$$X =_e Y \Rightarrow X =_\eta Y \Rightarrow X =_\zeta Y \Rightarrow X =_e Y .$$

To show that $X =_e Y$ implies $X =_\eta Y$, suppose that $XV =_\eta YV$ for all V. Then there is a variable $x \notin XY$ such that $Xx =_\eta Yx$. Then, by (η), we conclude

$$X =_\eta \lambda x.\, Xx =_\eta \lambda x.\, Yx =_\eta Y .$$

To show that $X =_\eta Y$ implies $X =_\zeta Y$, suppose that $x \notin X$. Then $(\lambda x.\, Xx)x = Xx$ without using any extensionality principle. By rule (ζ), we conclude that

$$\lambda x.\, Xx =_\zeta X .$$

Finally, the proof that $X =_\zeta Y$ implies $X =_e Y$ is similar to the corresponding proof for combinatory terms. This completes the theorem.

For the rest of the chapter we consider only combinatory terms, and we prove that extensional equality in the theory of combinators can be finitely axiomatized, i.e., there are a finite number of equations, the set of which we will call Ax, such that when the equations Ax are added to the postulates of Definition 2.19, the resulting equality is extensional.

Lemma 5.5. If a relation = for combinatory terms satisfies the clauses of Definition 2.19 as well as the following two equations:

Ax 1. $[x, y]\mathbf{S}(\mathbf{K}x)(\mathbf{K}y) = [x, y]\mathbf{K}(xy)$

Ax 2. $[x]\mathbf{S}(\mathbf{K}x)\mathbf{I} = \mathbf{I}$

then for all X, Y and variables x,

$$[x]\mathbf{XY} = \mathbf{S}([x]\mathbf{X})([x]\mathbf{Y}) .$$

Proof. The desired equality is an identity (by definition of the abstraction operation, Definition 2.10) unless either (i) $x \notin \mathbf{XY}$, or (ii) $x \notin \mathbf{X}$ and $\mathbf{Y} \equiv x$.

In case (i), $[x]\mathbf{XY} \equiv \mathbf{K}(\mathbf{XY}) = \mathbf{S}(\mathbf{KX})(\mathbf{KY})$ by Ax 1 $\equiv \mathbf{S}([x]\mathbf{X})([x]\mathbf{Y})$.

In case (ii), $[x]\mathbf{XY} \equiv \mathbf{X} = \mathbf{IX} = \mathbf{S}(\mathbf{KX})\mathbf{I}$ by Ax 2 $\equiv \mathbf{S}([x]\mathbf{X})([x]\mathbf{Y})$,

completing the proof.

By Theorem 5.4, it is sufficient to define Ax so that rule (ξ) is valid for the resulting equality. In particular, we must define Ax so that, for all X, Y, Z, and variables x, the following will be derivable.

(1) $[x]\mathbf{IX} = [x]\mathbf{X}$

(2) $[x]\mathbf{KXY} = [x]\mathbf{X}$

(3) $[x]\mathbf{SXYZ} = [x]\mathbf{XZ}(\mathbf{YZ})$.

If we assume Ax 1 and Ax 2, and, therefore, Lemma 5.5, the

following equations will be valid:

$$\begin{aligned}
&(4) \quad [x]\mathbf{I}X = \mathbf{S}(\mathbf{KI})([x]X)\ , \\
&(5) \quad [x]\mathbf{K}XY = \mathbf{S}([x]\mathbf{K}X)([x]Y) \\
&\qquad = \mathbf{S}(\mathbf{S}(\mathbf{KK})([x]X))([x]Y)\ , \\
&(6) \quad [x]\mathbf{S}XYZ = \mathbf{S}([x]\mathbf{S}XY)([x]Z) \\
&\qquad = \mathbf{S}(\mathbf{S}([x]\mathbf{S}X)([x]Y))([x]Z) \\
&\qquad = \mathbf{S}(\mathbf{S}(\mathbf{S}(\mathbf{KS})([x]X))([x]Y))([x]Z)\ , \\
&(7) \quad [x]XZ(YZ) = \mathbf{S}([x]XZ)([x]YZ) \\
&\qquad = \mathbf{S}(\mathbf{S}([x]X)([x]Z))(\mathbf{S}([x]Y)([x]Z))\ .
\end{aligned}$$

We see, then, that (1), (2), and (3) will follow from Ax 1, Ax 2, and the following.

Ax 3. $[x]\mathbf{S}(\mathbf{KI})x = [x]x$

Ax 4. $[x, y]\mathbf{S}(\mathbf{S}(\mathbf{KK})x)y = [x, y]x$

Ax 5. $[x, y, z]\mathbf{S}(\mathbf{S}(\mathbf{S}(\mathbf{KS})x)y)z = [x, y, z]\mathbf{S}(\mathbf{S}xz)(\mathbf{S}yz)$.

For example, (3) follows from (6), (7) and Ax 5, thus:

$$\begin{aligned}
[x]\mathbf{S}XYZ &= \mathbf{S}(\mathbf{S}(\mathbf{S}(\mathbf{KS})([x]X))([x]Y))([x]Z) \quad \text{by (6)} \\
&= ([x, y, z]\mathbf{S}(\mathbf{S}(\mathbf{S}(\mathbf{KS})x)y)z)([x]X)([x]Y)([x]Z) \\
&\qquad\qquad \text{by Theorem 2.16} \\
&= ([x, y, z]\mathbf{S}(\mathbf{S}xz)(\mathbf{S}yz))([x]X)([x]Y)([x]Z) \quad \text{by Ax 5} \\
&= \mathbf{S}(\mathbf{S}([x]X)([x]Z))(\mathbf{S}([x]Y)([x]Z)) \quad \text{by Theorem 2.16} \\
&= [x]XZ(YZ) \quad \text{by (7)}\ .
\end{aligned}$$

The proofs of (1) and (2) are similar.

Definition 5.6 (Axioms of extensionality). The set Ax of axioms of extensionality consists of the five equations Ax 1 through Ax 5. The relation $=_{ax}$ is defined by adding the equations of Ax to Definition 2.19.

Theorem 5.7 (Curry). $X =_e Y$ iff $X =_{ax} Y$.

Proof. Direct calculation readily shows that every equation of Ax is a consequence of rule (ξ), so that (by Theorem 5. 4), $X =_{ax} Y \Rightarrow X =_e Y$.

For the converse, it suffices (again by Theorem 5. 4) to show that rule (ξ) is valid for $=_{ax}$. That is, we show

(8) $\qquad X =_{ax} Y$

implies

(9) $\qquad [x]X =_{ax} [x]Y$.

This we do by induction on the length n of the proof of (8).

If $n \equiv 1$ and (8) is an instance of axiom schemes (I), (K), or (S), then (9) follows from (1), (2) or (3) respectively. These latter hold by virtue of Lemma 5. 5 and the discussion following it.

If $n \equiv 1$ and (8) is an equation of Ax, then $[x]X \equiv \mathbf{K}X$ and $[x]Y \equiv \mathbf{K}Y$, so (9) follows by rule (μ).

Suppose $n > 1$ and (8) is a consequence by rule (μ). Then $X \equiv ZU$, $Y \equiv ZV$, and $U =_{ax} V$ by a proof of length $n - 1$. By hypothesis of induction, $[x]U =_{ax} [x]V$, and by Lemma 5. 5,

$$[x]X =_{ax} \mathbf{S}([x]Z)([x]U), \quad [x]Y =_{ax} \mathbf{S}([x]Z)([x]V) .$$

Thus, (9) follows from the hypothesis of induction and rule (μ).

Rule (ν) is handled like rule (μ) and rules (σ) and (τ) are trivial. This completes the proof.

See [CLg.I] §6C for more discussion on the finite axiomatizability of extensional equality.

Exercises

1. Define the notion of simultaneous substitution for λ-terms. (This is assumed in Lemma 5. 3.)

2. Prove Lemma 5. 3.

3. Show $\mathbf{SK} =_{ax} \mathbf{KI}$ directly from Definition 5. 6.

6. The Equivalence of λ-Conversion and the Theory of Combinators

We have emphasized that combinators and λ-terms 'perform the same tasks', but we have observed (Remark 2.22) that the theories of Chapters 1 and 2, while similar, are not precisely equivalent. Here we show that when we consider extensional equality, there is an exact equivalence between the theories.

In this chapter, the equality symbol will always mean extensional equality, and we will make free use of the several equivalent ways of formulating this equality (Theorem 5.4). We will assume that the same sequence of variables is used in defining the combinatory terms as the λ-terms.

Definition 6.1 (The λ-transform and the H-transform).

(1) For any combinatory term X, the λ-transform of X, called X_λ, is defined inductively as follows:

(a) If X is an atom distinct from **S**, **K**, **I**, then $X_\lambda \equiv X$.

(b) $\mathbf{I}_\lambda \equiv \lambda u.\, u$

(c) $\mathbf{K}_\lambda \equiv \lambda uv.\, u$

(d) $\mathbf{S}_\lambda \equiv \lambda uvw.\, uw(vw)$

(e) If $X \equiv YZ$, then $X_\lambda \equiv Y_\lambda Z_\lambda$.

(Here, u, v, w are the first three in the list of variables.)

(2) For any λ-term X, the H-transform of X, called X_H, is defined inductively as follows:

(a) If X is an atom, then $X_H \equiv X$.

(b) If $X \equiv YZ$, then $X_H \equiv Y_H Z_H$.

(c) If $X \equiv \lambda x Y$, then $X_H \equiv [x]Y_H$.

Note that for combinatory terms X, X_λ is a λ-term, and that for λ-terms M, M_H is a combinatory term. Also the same variables occur free in X_λ as occur in X, and the same variables occur in M_H as occur free in M.

Lemma 6.2. *Let X, Y be λ-terms, and x a variable. Then*

(a) $([Y/x]X)_H \equiv [Y_H/x]X_H$,

(b) *if* X *and* Y *are congruent, then* $X_H \equiv Y_H$.

Proof. The proofs are by induction on the number of λ's in X. If X has no λ's, both results are clear, so we need only consider the case that $X \equiv \lambda y. X'$, and the hypothesis of induction applies to X'.

To prove (a) we may assume that x is free in X. If y is not free in Y, then

$$[Y/x]X \equiv \lambda y. [Y/x]X'$$

so that

$$\begin{aligned}([Y/x]X)_H &\equiv [y]([Y/x]X')_H \\ &\equiv [y]([Y_H/x]X'_H), \text{ by hypothesis of induction} \\ &\equiv [Y_H/x][y]X'_H, \text{ by Lemma 2.17} \\ &\equiv [Y_H/x]X_H .\end{aligned}$$

If y is free in Y, then for some variable z which is not free in YX',

$$[Y/x]X \equiv \lambda z. [Y/x][z/y]X' .$$

Note that the hypothesis of induction applies to [z/y]X', so that we get the following identities:

$$\begin{aligned}([Y/x]X)_H &\equiv [z]([Y/x][z/y]X')_H \\ &\equiv [z][Y_H/x][z/y]X'_H, \text{ by hyp. of induc. twice} \\ &\equiv [Y_H/x][z][z/y]X'_H, \text{ by Lemma 2.17}\end{aligned}$$

$$\equiv [Y_H/x][y]X'_H, \quad \text{by Lemma 2.17}$$
$$\equiv [Y_H/x]X_H .$$

To prove (b), it is enough to consider the case when Y is the result of only one change of bound variable in X (and $X \equiv \lambda y. X'$). If this change takes place inside X', then

$$Y \equiv \lambda y. Y'$$

with Y' congruent to X', and (b) follows by the induction hypothesis.

If the change is not in X', then we must have

$$X \equiv \lambda y. X' , \qquad Y \equiv \lambda z. [z/y]X' ,$$

with y not bound in X' and z neither free nor bound in X'. Then

$$X_H \equiv [y]X'_H \equiv [z][z/y]X'_H , \quad \text{by Lemma 2.17}$$
$$\equiv [z]([z/y]X')_H, \quad \text{by part (a)}$$
$$\equiv Y_H .$$

__Corollary 6.2.1.__ __If__ X __and__ Y __are__ λ-__terms__, __then__

$$((\lambda x. X)Y)_H = ([Y/x]X)_H .$$

Proof.

$$((\lambda x. X)Y)_H \equiv (\lambda x. X)_H Y_H$$
$$\equiv ([x]X_H)Y_H$$
$$= [Y_H/x]X_H, \quad \text{by Theorem 2.12}$$
$$\equiv ([Y/x]X)_H .$$

We are now ready for the equivalence theorem.

__Theorem 6.3.__

(a) __For combinatory terms__ X, Y, $X = Y$ __iff__ $X_\lambda = Y_\lambda$.

(b) __For__ λ-__terms__ X, Y, $X = Y$ __iff__ $X_H = Y_H$.

(c) __For combinatory terms__ X, $X_{\lambda H} \equiv X$.

(d) __For__ λ-__terms__ X, $X_{H\lambda} = X$.

Remark. Notice that (c) is an identity, but (d) is only an equality. By (c), the λ-transform is a one-to-one mapping from combinatory terms into the set of λ-terms; but the H-transform is only one-to-one if we consider equality-classes of terms instead of the terms themselves.

Proof of Theorem. The proof is in six parts.

(1) Prove the 'only if' implication of (a), by induction on the length of proof of $X = Y$ in the formulation of equality with rule (ξ). This is straightforward.

(2) Prove the 'only if' implication of (b), by induction on the length of proof of $X = Y$, again using rule (ξ). If $X = Y$ is an axiom by (α) of Definition 1.9, then $X_H = Y_H$ by Lemma 6.2(b). If $X = Y$ is an axiom by (β) of Definition 1.9, then $X_H = Y_H$ by Corollary 6.2.1. The other cases are immediate.

(3) Prove (c), by induction on the number of occurrences of atoms in X. This is easy.

(4) Prove (d), by induction on X. The proof reduces to the case $X \equiv \lambda x. Y$, with $Y_{H\lambda} = Y$ assumed by the hypothesis of induction. If we can prove that for all Z,

$$\text{(i)} \qquad ([x]Z)_\lambda = \lambda x. Z_\lambda ,$$

then we shall have in the above case,

$$\begin{aligned} X_{H\lambda} \equiv ([x]Y_H)_\lambda &= \lambda x. Y_{H\lambda} \\ &= \lambda x. Y, \quad \text{by induction hypothesis} \\ &\equiv X . \end{aligned}$$

So it is enough to prove (i). This is done by induction on Z, as follows.

If $Z \equiv x$:

$$([x]Z)_\lambda \equiv I_\lambda \equiv \lambda u. u = \lambda x. x \equiv \lambda x. Z_\lambda .$$

If $x \notin Z$:

$$([x]Z)_\lambda \equiv (\mathbf{K}Z)_\lambda$$
$$\equiv (\lambda uv.\, u)Z_\lambda$$
$$= (\lambda yx.\, y)Z_\lambda$$
$$= \lambda x.\, Z_\lambda\ .$$

If $Z \equiv UV$, then by Lemma 5. 5,

$$[x]Z = \mathbf{S}([x]U)([x]V)\ .$$

Hence by part (1) of the proof,

$$([x]Z)_\lambda = \mathbf{S}_\lambda([x]U)_\lambda([x]V)_\lambda$$
$$= \mathbf{S}_\lambda(\lambda x.\, U_\lambda)(\lambda x.\, V_\lambda), \text{ by induc. hyp.}$$
$$= (\lambda yzx.\, yx(zx))(\lambda x.\, U_\lambda)(\lambda x.\, V_\lambda)$$
$$= \lambda x.\, ((\lambda x.\, U_\lambda)x)((\lambda x.\, V_\lambda)x)$$
$$= \lambda x.\, U_\lambda V_\lambda$$
$$\equiv \lambda x.\, Z_\lambda\ .$$

This completes (i), and hence (4).

(5) Prove the 'if' implication of (a). Assume $X_\lambda = Y_\lambda$. By part (2) of the proof, $X_{\lambda H} = Y_{\lambda H}$. By (c), the last equation is the desired result.

(6) Prove the 'if' implication of (b). Assume $X_H = Y_H$. By (a), $X_{H\lambda} = Y_{H\lambda}$. By (d), $X = Y$ follows.

7. Strong Reduction

In Chapters 1 and 2, reduction relations were introduced to represent the process of calculating the value of a function at a given argument (Definition 1. 9 and Definition 2. 3), and only later were these relations extended to equalities (Definition 1. 11 and Definition 2. 19). The obvious existence of irreducible terms and the Church-Rosser theorems (Theorem 1. 13 and Theorem 2. 21) resulted in consistency proofs for the two theories of equality, and a criterion for proving the inequality of terms (Corollary 1. 13. 3 and Corollary 2. 21. 3).

It seemed natural in Chapter 5 to define extensional equality without first defining corresponding reduction relations. Such relations are the subject of this chapter. We shall refer only briefly to the relation for λ-terms, confining our attention almost exclusively to combinatory terms and their reduction relation, strong reduction. After the definition of strong reduction, we shall consider three topics: the Church-Rosser theorem, the 'linearization' or axiomatization of strong reduction, and the characterization of the irreducible terms.

In this chapter, equality of terms will always mean extensional equality, and 'term , unless specified to the contrary, will mean <u>combinatory</u> term. The reduction relation $X \rhd Y$ of Definition 2. 3 will be referred to as <u>weak</u> reduction.

Definition 7. 1 <u>(Strong reduction)</u>. The relation $X \succ\!\!- Y$ (X strongly reduces to Y) is defined by induction as follows:

<u>Axiom-schemes:</u>

(**I**) $\mathbf{I}X \succ\!\!- X$ for all X ;

(**K**) $\mathbf{K}XY \succ\!\!- X$ for all X, Y ;

(S) $\mathbf{S}XYZ \succ XZ(YZ)$ for all X, Y, Z ;

(ρ) $X \succ X$ for all X.

Deduction-rules:

(μ) $X \succ X'$ implies $ZX \succ ZX'$;

(ν) $X \succ X'$ implies $XZ \succ X'Z$;

(τ) $X \succ Y$ and $Y \succ Z$ implies $X \succ Z$;

(ξ) $X \succ Y$ implies $[x]X \succ [x]Y$.

We say that X is (strongly) irreducible iff $X \succ Y$ implies that $Y \equiv X$.

Lemma 7. 2.

(a) If $X \rhd Y$, then $X \succ Y$.

(b) If $X \succ Y$, then $X = Y$.

(c) If $X \succ Y$ and $x \notin X$, then $x \notin Y$.

(d) A term containing no occurrences of combinators is irreducible.

Proof. Parts (a), (b), (c) are immediate from the definition of strong reduction. For part (d), note first that if [x]X contains no combinators, then neither does X; and then show, by induction on the length of the proof of $X \succ Y$, that if $Y \not\equiv X$, then X contains at least one occurrence of a combinator.

The converses of Lemma 7. 2 (a) and (b) are false, as the following examples show:

(a) $\mathbf{SK} \equiv [x, y]\mathbf{SK}xy \succ [x, y]\mathbf{K}y(xy) \succ [x, y]y \equiv \mathbf{KI}$;

(b) $x = \mathbf{I}x$ but not $x \succ \mathbf{I}x$, by Lemma 7. 2(d).

To prove the Church-Rosser Theorem for strong reduction we must return briefly to λ-conversion.

Definition 7. 3 (Strong reduction for λ-terms; Curry's $\beta\eta$-reduction). The relation $X \succ_{\lambda} Y$ between λ-terms X, Y is got by adding the following axiom-scheme (η) to the axiom-schemes and rules of reduction in Definition 1. 9:

(η) $\lambda x. Xx \succ_{\lambda} X$, provided that $x \notin X$.

Lemma 7.4. <u>For λ-terms</u> X, Y; <u>if</u> $X \succ_{\lambda} Y$, <u>then</u> $X_H \succ Y_H$.

Proof. The proof by induction on the length of proof of $X \succ_{\lambda} Y$ is easy after observing that the proof of Corollary 6.2.1 actually shows that

$$((\lambda x. X)Y)_H \rhd ([Y/x]X)_H .$$

Theorem 7.5 (Church-Rosser). <u>For λ-terms</u> X, Y; <u>if</u> $X = Y$ <u>then there is a</u> Z <u>such that</u> $X \succ_{\lambda} Z$ <u>and</u> $Y \succ_{\lambda} Z$.

Proof. See Appendix 1, Remark 2. The Church-Rosser Theorem for strong reduction is then a simple corollary of Theorem 7.5, Lemma 7.4, and Theorem 6.3(a), (c), as follows:

Theorem 7.6 (Church-Rosser). <u>For combinatory terms</u> X, Y; <u>if</u> $X = Y$ <u>then there is a</u> Z <u>such that</u> $X \succ Z$ <u>and</u> $Y \succ Z$.

Proof. If $X = Y$, then $X_\lambda = Y_\lambda$ by Theorem 6.3(a). By Theorem 7.5 there is a U such that $X_\lambda \succ_{\lambda} U$ and $Y_\lambda \succ_{\lambda} U$. Then $X \equiv X_{\lambda H} \succ U_H$ and $Y \equiv Y_{\lambda H} \succ U_H$, by Theorem 6.3(c) and Lemma 7.4. We may then take Z to be U_H.

The above definition of strong reduction is the original one of Curry [CLg. I] §6F1; it makes easy a comparison with λ-conversion and the proof of the Church-Rosser Theorem. Because of rule (ξ), however, strong reduction as formulated above does not have the linear character of weak reduction (or of reduction of λ-terms), whereby $X \rhd Y$ implies that X can be carried into Y by a sequence of replacements of redexes by contracta.

However, it is possible to define redexes and contracta so that strong reduction can be linearized. What we shall do is define axioms of strong reduction (whose left and right hand sides will be the redexes and contracta, respectively), and the relation $\succ\!\!-_a$ generated by them; and then we shall prove the equivalence of the relations $\succ\!\!-$ and $\succ\!\!-_a$.

The axioms are presented in terms of an infinite system of axiom schemes which are generated from a finite list of basic axiom schemes by a generating rule (GR). The rule (GR) is designed precisely to insure the validity of rule (ξ) for $\succ\!\!-_a$.

Because there are infinitely many axiom schemes, and because of the complexity of the abstraction operation which is involved in rule (GR), it is necessary to be more systematic than we have been up to now in stating axiom schemes, so that we may analyze their structure. To this end, we shall introduce the meta-variables, new atoms which identify the places at which substitutions are to be made in the axiom schemes.

Definition 7.7 Meta-variables and term schemes.

(a) Let A, B, C be three new atoms, which we call meta-variables.

(b) Term schemes are defined by induction as follows: the variables, the meta-variables, the combinators **I, K, S** and any other constant atomic terms, are (atomic) term schemes; if P, Q are term schemes, then (PQ) is a term scheme.

The notion of term is unchanged, so the terms are the term schemes without meta-variables.

Notation. (a) The letters 'M', 'N', 'P', 'Q', 'R' will stand for term schemes in this chapter. The letters 'U', 'V', 'W', 'X', 'Y', 'Z' will stand for terms.

(b) We shall abbreviate the result of substituting terms for meta-variables: 'R[M, N, P]' will stand for [M/A][N/B][P/C]R.

(c) Choose a fixed variable x. For any term scheme P such that $x \notin P$, define

$$P^1 \equiv P, \quad P^2 \equiv Px, \quad P^3 \equiv x .$$

(d) Let P be any term scheme such that $x \notin P$. Define

$$P_{ijk} \equiv P[A^i, B^j, C^k]. \qquad (1 \leq i, j, k \leq 3)$$

(e) We extend the definition of $[x]M$ to the case that M is a term scheme simply by allowing M, U, V in Definition 2.10 to be arbitrary term schemes.

We are now ready for the main definitions.

Definition 7.8 <u>(Axiom schemes of strong reduction)</u>. The axiom schemes of strong reduction are defined by induction as follows:

(1) $\mathbf{I}A \succ_a A$

(2) $\mathbf{K}AB \succ_a A$

(3) $\mathbf{S}ABC \succ_a AC(BC)$

(4) $\mathbf{S}(\mathbf{K}A)\mathbf{I} \succ_a A$

(5) $\mathbf{S}(\mathbf{K}A)(\mathbf{K}B) \succ_a \mathbf{K}(AB)$

(GR) If $P \succ_a Q$ is an axiom scheme, then for any i, j, k,

$$[x]P_{ijk} \succ_a [x]Q_{ijk}$$

is an axiom scheme.

We call P the <u>redex scheme</u>, and Q the <u>contractum scheme</u> of the axiom scheme $P \succ_a Q$.

Definition 7.9. We define the relation $\succ_a$ between combinatory terms as follows:

(a) If $P \succ_a Q$ is an axiom scheme, and if there are terms U, V, W such that $X \equiv P[U, V, W]$ and $Y \equiv Q[U, V, W]$, then $X \succ_a Y$;

(ρ) $X \succ_a X$;

(μ) If $X \succ_a Y$, then $ZX \succ_a ZY$;

(ν) If $X \succ_a Y$, then $XZ \succ_a YZ$;

(τ) If $X \succ_a Y$ and $Y \succ_a Z$, then $X \succ_a Z$.

If $X \succ_a Y$ by (a), we say that X is a redex with contractum Y and that $X \succ_a Y$ is an axiom of strong reduction.

Lemma 7.10. $X \succ_a Y$ iff there is a sequence of terms $X \equiv X_1, X_2, \ldots, X_n \equiv Y$ such that each X_{i+1} results from X_i by the replacement of a single component which is a redex by its contractum.

Remark 7.11. (i) The choice of the variable x for the generating rule (GR) is arbitrary, so that, for example, if P is a redex scheme and $X \equiv ([x]P_{ijk})[U, V, W]$, we may assume that $x \notin UVW$.

(ii) From axiom schemes (iv), (v), it follows that for any X and Y, $\mathbf{S}([x]X)([x]Y) \succ_a [x]XY$. (Cf. Lemma 5.5.)

(iii) If we call a redex X significant when it has a contractum $Y \not\equiv X$, and if we call an axiom $X \succ_a Y$ significant when $Y \not\equiv X$, then the notions of significant redex and significant axiom are decidable, as is shown in Lercher [DHA] Theorem 3.

Lemma 7.12. If $x \notin PUVW$, then

$([x]P_{ijk})[U, V, W] \equiv [x](P[U^i, V^j, W^k])$.

Proof. First note that for any Q, if $x \notin UVW$, then $([x]Q)[U, V, W] \equiv [x](Q[U, V, W])$ by Lemma 2.17(b). Then note that $P_{ijk}[U, V, W] \equiv P[U^i, V^j, W^k]$.

Corollary 7.12.1. If $X \succ_a Y$, then $X \succ Y$.

Proof. First suppose that $X \succ_a Y$ is an axiom and that $X \equiv P[U, V, W]$, $Y \equiv Q[U, V, W]$ for an axiom scheme $P \succ_a Q$. Use induction on the number n of applications of (GR) in deriving that $P \succ_a Q$ is an axiom scheme. If $n \equiv 0$, we calculate directly. For example, Let $P \equiv \mathbf{S}(\mathbf{K}A)(\mathbf{K}B)$, $Q \equiv \mathbf{K}(AB)$, and choose $x \notin UV$. Then

$$X \equiv [x]\mathbf{S}(\mathbf{K}U)(\mathbf{K}V)x \succ [x]\mathbf{K}Ux(\mathbf{K}Vx)$$
$$\succ [x]UV \equiv Y .$$

Now suppose $n > 0$ and $P \equiv [x]P'_{ijk}$, $Q \equiv [x]Q'_{ijk}$ for an axiom scheme $P' \succ_a Q'$. Let $X' \equiv P'[U^i, V^j, W^k]$ and $Y' \equiv Q'[U^i, V^j, W^k]$, so that $X' \succ Y'$ by hypothesis of induction. By the lemma, $X \equiv [x]X' \succ [x]Y' \equiv Y$.

This proves the corollary for axioms. Since the rules (μ), (ν) and (τ) for $\succ_a$ are also valid for $\succ$, the proof is complete by induction on the length of the proof of $X \succ_a Y$.

Lemma 7.13. Suppose $x \notin P$. Then for any terms U, V, W there are terms U', V', W' and integers $1 \leq i, j, k \leq 3$ such that

$$[x](P[U, V, W]) \equiv ([x]P_{ijk})[U', V', W'] .$$

Hence, if X is any redex, $[x]X$ is also a redex.

Proof. Define U' and i as follows: if $x \notin U$, then $U' \equiv U$ and $i \equiv 1$; if $x \in U \not\equiv x$, then $U' \equiv [x]U$ and $i \equiv 2$; if $U \equiv x$, then U' is arbitrary and $i \equiv 3$. Similarly for V' and j, and W' and k.

We now proceed by induction on the structure of $P[U, V, W]$. The indices i, j, k have been chosen so that the evaluation of $[x]P_{ijk}$ parallels that of $[x](P[U, V, W])$ as closely as possible; if $x \notin U$, then $x \notin A^i$, if $U \equiv x$, then $A^i \equiv x$, etc.

If $x \notin P[U, V, W]$, then in the above definition the index of every meta-variable in P is 1, that is $i \equiv j \equiv k \equiv 1$, so $P_{ijk} \equiv P$. Hence,

$$[x](P[U, V, W]) \equiv \mathbf{K}(P[U, V, W]) \equiv (\mathbf{K}P)[U, V, W] \equiv ([x]P_{ijk})[U', V', W'] .$$

If $x \equiv P[U, V, W]$, then P is a meta-variable whose index is 3 (e.g., $P \equiv A$ and $U \equiv x$, so that $i \equiv 3$). Then

$$[x](P[U, V, W]) \equiv \mathbf{I} \equiv ([x]P_{ijk})[U', V', W'] .$$

If $P[U, V, W] \equiv Xx$ with $x \notin X$, then $P \equiv QR$ where R is a meta-variable with index 3 and the index of every meta-variable in Q is 1. Hence, $P_{ijk} \equiv Qx$ and $[x]P_{ijk} \equiv Q$, so that

$$[x](P[U, V, W]) \equiv X \equiv Q[U, V, W] \equiv ([x]P_{ijk})[U', V', W'] .$$

If $P[U, V, W] \equiv XY$ and either $x \in X$ or $x \in Y \not\equiv x$, then either P is a meta-variable (e.g. $P \equiv A$) and $U \equiv XY$, or $P \equiv QR$ and $X \equiv Q[U, V, W]$ and $Y \equiv R[U, V, W]$. In the first case we have

$$U' \equiv [x]U, \qquad i \equiv 2$$

and so

$$\begin{aligned}[x](P[U, V, W]) \equiv [x]U \equiv U' &\equiv ([x]Ax)[U', V', W'] \\ &\equiv ([x]P_{ijk})[U', V', W'] .\end{aligned}$$

In the second case, by definition of i, j, k and A^i, B^j, C^k, we have $x \in Q_{ijk}$ or $x \in R_{ijk} \not\equiv x$. The hypothesis of induction applies to Q and R, so that

$$\begin{aligned}[x](P[U, V, W]) &\equiv \mathbf{S}([x]X)([x]Y) \\ &\equiv \mathbf{S}([x](Q[U, V, W]))([x](R[U, V, W])) \\ &\equiv \mathbf{S}(([x]Q_{ijk})[U', V', W'])(([x]R_{ijk})[U', V', W']) \\ &\equiv ([x]P_{ijk})[U', V', W'] .\end{aligned}$$

Corollary 7.13.1. <u>If $X \succ_a Y$ is an axiom, then $[x]X \succ_a [x]Y$ is also an axiom.</u>

Proof. Let $P \succ_a Q$ be an axiom scheme such that $X \equiv P[U, V, W]$ and $Y \equiv Q[U, V, W]$. Choose U', V', W', i, j, k as in the lemma. Then

$$[x]P_{ijk} \succ_a [x]Q_{ijk}$$

is an axiom scheme such that

$$([x]P_{ijk})[U', V', W'] \succ_a ([x]Q_{ijk})[U', V', W']$$

is an axiom. The lemma gives the conclusion immediately.

This completes the preliminary work for the equivalence theorem.

Theorem 7.14. *For any terms* X, Y, $X \succ_a Y$ *iff* $X \succ Y$.

Proof. That $X \succ_a Y$ implies $X \succ Y$ is Corollary 7.12.1. For the converse we use an induction on the proof of $X \succ Y$. All the postulates for strong reduction except for Rule (ξ) are also postulates for $\succ_a$, so that we need only to show that

(i) $\quad X \succ_a Y$

implies

(ii) $\quad [x]X \succ_a [x]Y$.

The proof is by induction on the proof of (i).

If (i) is an axiom, then (ii) follows by Corollary 7.13.1. If (i) is a consequence of Rule (μ), then $X \equiv ZU$ and $Y \equiv ZV$ and $U \succ_a V$. The hypothesis of induction is that $[x]U \succ_a [x]V$. There are three cases.

(a) Suppose $x \notin U$. Then by Corollary 7.12.1 and Lemma 7.2(c), $x \notin V$. If $x \notin Z$, then $[x]X \equiv \mathbf{K}(ZU)$ and $[x]Y \equiv \mathbf{K}(ZV)$, so that (ii) follows by Rule (μ). If $x \in Z$, then

$[x]X \equiv \mathbf{S}([x]Z)(\mathbf{K}U)$ and $[x]Y \equiv \mathbf{S}([x]Z)(\mathbf{K}V)$, so that, again, (ii) follows by (μ).

(b) Suppose $U \equiv x$. Then Corollary 7.12.1 and Lemma 7.2(d) imply that $V \equiv x$. Then (ii) is an instance of (ρ).

(c) Suppose $x \in U \not\equiv x$. Then $[x]X \equiv \mathbf{S}([x]Z)([x]U)$. By the hypothesis of induction and Rule (μ), we have

$$[x]X \succ_a \mathbf{S}([x]Z)([x]V) .$$

But Remark 7.11(ii) says that

$$\mathbf{S}([x]Z)([x]V) \succ_a [x]ZV \equiv [x]Y .$$

Thus, (ii) follows in this case.

If (i) is a consequence of an inference by Rule (ν), the proof is similar. Since Rule (τ) is trivial, we are done.

Remark 7.15. The definition of the axiom schemes, Definition 7.8, allows for some unnecessary axiom schemes and axioms. For example, if $P \succ_a Q$ is an axiom scheme, then, setting $i \equiv j \equiv k \equiv 1$ in (GR) we get that $\mathbf{K}P \succ_a \mathbf{K}Q$ is an axiom scheme. But any axiom

$$\mathbf{K}P[U, V, W] \succ_a \mathbf{K}Q[U, V, W]$$

follows from the axiom

$$P[U, V, W] \succ_a Q[U, V, W]$$

by Rule (μ). Hence, the elimination of $\mathbf{K}P \succ_a \mathbf{K}Q$ as an axiom scheme does not alter the relation $\succ_a$. Other unnecessary axiom schemes are

$$[x]\mathbf{S}ABx \succ_a [x]Ax(Bx) ,$$
$$[x]\mathbf{K}Ax \succ_a [x]A ,$$

$$[x]\mathbf{I}x \succ_a [x]x ,$$

in all of which the contractum scheme is identical to the redex scheme. Finally, if we add a new axiom scheme

$$(6) \qquad \mathbf{S}(\mathbf{KI}) \succ_a \mathbf{I} ,$$

we need never apply (GR) to the axiom scheme $\mathbf{I}A \succ_a A$. For, in the only case not already considered, the case $i \equiv 2$, the result of applying (GR) to the scheme $\mathbf{I}A \succ_a A$ is

$$\mathbf{S}(\mathbf{KI})A \succ_a A ,$$

which follows from schemes (6) and (1), thus:

$$\mathbf{S}(\mathbf{KI})A \succ_a \mathbf{I}A \succ_a A .$$

We are then led to the following modification of the axiom schemes for strong reduction.

Definition 7.16 (modifying Definition 7.8). The modified set of axiom schemes for strong reduction is defined by induction as follows:

(1) $\mathbf{I}A \succ_a A,$

(2) $\mathbf{K}AB \succ_a A,$

(3) $\mathbf{S}ABC \succ_a AC(BC),$

(4) $\mathbf{S}(\mathbf{K}A)\mathbf{I} \succ_a A,$

(5) $\mathbf{S}(\mathbf{K}A)(\mathbf{K}B) \succ_a \mathbf{K}(AB),$

(6) $\mathbf{S}(\mathbf{KI}) \succ_a \mathbf{I} ,$

(GR) If $P \succ_a Q$ is an axiom scheme, then so is $[x]P_{ijk} \succ_a [x]Q_{ijk}$, in all except the following cases:

(a) $i \equiv j \equiv k \equiv 1$, or $i \equiv j \equiv 1$ and $C \notin PQ$, or $i \equiv 1$ and $B, C \notin PQ$;

(b) $P \succ_a Q$ is scheme (3) with $i \equiv j \equiv 1$ and $k \equiv 3$, or scheme (2) with $i \equiv 1$ and $j \equiv 3$;

(c) $P \succ_a Q$ is scheme (1) or (6).

From now on, the definitions of redex scheme, contractum scheme, redex, contractum, axiom, and the relation $\succ_a$ will refer to this modified set of schemes instead of to Definition 7.8. It can be shown that the properties of being a modified redex scheme, an axiom scheme, a redex, and an axiom, are effectively decidable. Also, each redex has only a finite number of contracta, and there is an effective method for generating all of these. (Lercher [DHA] Theorem 3.)

Theorem 7.17. The relation $\succ_a$ defined by the modified set of axioms is still the same as $\succ$.

Proof. See Remark 7.15.

Now let us turn to the important problem of giving a simple description of the strongly irreducible terms. This turns out to be fairly difficult, in contrast to weak reduction and λ-reduction, because of the infinite number of axiom schemes involved. The results will be merely stated here without proof.

Definition 7.18. A combinatory term X is in (strong) normal form iff either

(i) X is an atom distinct from **S**, **K** and **I**, or

(ii) $X \equiv aX_1 \ldots X_n$ for terms $X_1, \ldots, X_n$ in normal form and an atom a distinct from **S**, **K** and **I**, or

(iii) $X \equiv [x]Y$ for a term Y in normal form, and a variable x.

If a term U strongly reduces to an X in normal form, we say that X is a (strong) normal form of U.

Theorem 7.19. For combinatory terms X, the following are equivalent:

(a) X is strongly irreducible,

(b) X contains no redexes (in the sense of Definition 7.16),

(c) X is in strong normal form.

Proof. That every term in normal form contains no redexes, is proved in Hindley and Lercher [SPC]. The fact that every term containing no redexes is irreducible is obvious. Finally, Lercher [SRN] Theorem 3.2 shows that every irreducible term is in normal form. (Alternative proofs of these results are given in [CLg. II] §11E.)

By this theorem, we have in Definition 7.18 quite a simple characterization of the set of irreducible terms; and in Lercher [SRN] §5, it is shown that this set is decidable. This set also ties in very closely with normal forms in λ-conversion. Suppose we say that a λ-term X is in <u>strong</u> λ-<u>normal form</u> iff no component of X has form

$$(\lambda x. M)N, \qquad (\lambda x. Ux) \qquad (x \notin U) .$$

(This is the analogue for $\succ_{\lambda}$ of the definition of normal form for non-extensional λ-reduction, Definition 1.7.) Then we have the following easy result.

Theorem 7.20. *A combinatory term X is in strong normal form iff $X \equiv V_H$ for some λ-term V in strong λ-normal form.*

Remark 7.21. Notice incidentally that Theorem 7.19 would fail if the combinator **I** was defined as **SKK** instead of being taken as an atom. Because if **I** ≡ **SKK**, we would have

$$\mathbf{I} \equiv [x]x \text{ in normal form,}$$

$$\begin{aligned} \mathbf{I} \equiv \mathbf{SKK} &\succ \mathbf{KIK} \text{ since } \mathbf{SK} \succ \mathbf{KI}, \\ &\equiv \mathbf{K(SKK)K} \\ &\succ \mathbf{K(KIK)K}, \text{ etc.} \end{aligned}$$

Finally, the following lemma will be used in Chapter 9. It is easy to see that every combinatory term X can be expressed in the form

$$X \equiv aX_1 \dots X_n$$

where a is an atom and $n \geq 0$. (Use induction on X.) This occurrence of a is called the head of X.

Lemma 7.22. If $X \succ Y$ and X has form $aX_1 \dots X_n$, where a is an atom not a combinator, then Y has form $aY_1 \dots Y_n$, where $X_i \succ Y_i$ for $i = 1, \dots, n$.

Proof. By induction on Definition 7.16, every redex scheme is headed by a combinator. The result then follows from Lemma 7.10.

Exercises

1. Show that every strongly irreducible combinatory term is weakly irreducible, but not vice versa.

2. Show that a combinatory term may have a strong normal form but not have a weak normal form. (Contrast Ex. 1.)

3. Show that a term may have a weak normal form but no strong normal form.

8. Combinators with Types: First Approach

In Remark 2.23 it was pointed out that combinators cannot be easily interpreted as functions in set theory. In this chapter we shall look at a modified system whose terms can easily be interpreted as functions. It has proved useful as a basis for defining certain classes of functions in proof-theory (see Chapter 11), and its λ-analogue has been used for logical type-theories (see references in Chapter 10).

Before defining the modified combinatory terms, we shall first define a set of types. To do this, we assume that there are given certain basic types, each of which is intended to represent a particular set. For example, there may be given a basic type **N** to represent the set of all natural numbers.

Definition 8.1 (Types).

(i) Each basic type is a type.

(ii) If α and β are types, then $(\mathbf{F}\alpha\beta)$ is a type.

We assume that the types with form $(\mathbf{F}\alpha\beta)$ are distinct from the basic types, and that $(\mathbf{F}\alpha\beta) \equiv (\mathbf{F}\alpha'\beta')$ implies $\alpha \equiv \alpha'$ and $\beta \equiv \beta'$.

A type with the form $(\mathbf{F}\alpha\beta)$ (read as 'functions from α into β') is intended to represent a set of functions from the set represented by α into the set represented by β. The exact set of functions which it denotes will depend on the context where the typed combinators are used. When this is specified, each type comes to represent a set of individuals or functions.

Notation. Greek letters will denote types, and a type $(\mathbf{F}\alpha\beta)$ will often be written as $\mathbf{F}\alpha\beta$. We shall sometimes use the abbreviation

$$\mathbf{F}_n\alpha_1\ldots\alpha_n\beta \equiv \mathbf{F}\alpha_1(\mathbf{F}\alpha_2(\ldots(\mathbf{F}\alpha_n\beta)\ldots)) .$$

Incidentally, the '**F**' plays no essential role in Definition 8.1 and could well be omitted here; but it will play a role in Chapter 10, where other logical constants can be defined in terms of it. Common alternative notations for $(\mathbf{F}\alpha\beta)$ are $(\alpha \rightarrow \beta)$, $(\alpha\beta)$ and $(\beta\alpha)$.

Now, in order to define the typed terms, we shall assume that for each type α there exists an infinite of variables v^α (and $\alpha \not\equiv \beta \Rightarrow v^\alpha \not\equiv v^\beta$), and a distinct atom $\mathbf{I}_\alpha$; for each ordered pair α, β there is a distinct atom $\mathbf{K}_{\alpha\beta}$, and for each triple α, β, γ there is a distinct atom $\mathbf{S}_{\alpha\beta\gamma}$. There may also be some constant atoms c^δ, each with its own associated type, δ.

Definition 8.2 (Typed combinatory terms).

(i) Each variable v^α is a term, with type α; each constant atom c^δ is a term, with type δ; all the $\mathbf{I}_\alpha$, $\mathbf{K}_{\alpha\beta}$, $\mathbf{S}_{\alpha\beta\gamma}$ are terms (called atomic combinators), and

$\mathbf{I}_\alpha$ has type $\mathbf{F}\alpha\alpha$,
$\mathbf{K}_{\alpha\beta}$ has type $\mathbf{F}\alpha(\mathbf{F}\beta\alpha)$,
$\mathbf{S}_{\alpha\beta\gamma}$ has type $\mathbf{F}(\mathbf{F}\alpha(\mathbf{F}\beta\gamma))(\mathbf{F}(\mathbf{F}\alpha\beta)(\mathbf{F}\alpha\gamma))$.

(These types will be motivated in Remark 8.3.)

(ii) If $\mathbf{X}$ is a term with type $\mathbf{F}\alpha\beta$, and $\mathbf{Y}$ is a term with type α, then $(\mathbf{XY})$ is a term with type β.

The same assumptions and notation conventions about terms will be made as after Definitions 2.1 and 1.1. It can be seen from the above definition that no term has more than one type; sometimes its type will be written as a superscript, for example

$$v^{\mathbf{F}\alpha\beta}, \quad \mathbf{X}^\alpha .$$

Each term **X** with a type α is intended to represent a member of the set represented by α. Under this interpretation, rule (ii) above is valid because if **X** represents a function $\phi_{\mathbf{X}}$ from α into β, and **Y** represents a member $\phi_{\mathbf{Y}}$ of α, then **XY** represents a member of β, namely the result of applying $\phi_{\mathbf{X}}$ to $\phi_{\mathbf{Y}}$.

The constant atoms could perhaps include the $\bar{0}$ and $\bar{\sigma}$ from Remark 3.10; in this case it would be natural to give them the types **N** and **FNN** respectively (if **N** was the type of the natural numbers).

Remark 8.3 (The types of the combinators). To explain these types in Definition 8.2, imagine that we are assigning types to the combinatory terms of Chapter 2, beginning with the atoms and using rule (ii), Definition 8.2, for composite terms.

For **I**, we have **IX** = **X**, and equal terms should have the same interpretation. Hence if **X** has a type α, the term **IX** should have the same type. This can be achieved by giving **I** the type $\mathbf{F}\alpha\alpha$, since then **IX** has type α by rule (ii) above. But **IX** = **X** holds for every **X**, so we must either give **I** an infinity of types, $\mathbf{F}\alpha\alpha$ for all α, or else replace **I** by an infinity of combinators $\mathbf{I}_\alpha$, each with one type $\mathbf{F}\alpha\alpha$ and a restricted axiom-scheme

$$\mathbf{I}_\alpha \mathbf{X}^\alpha = \mathbf{X}^\alpha .$$

The first alternative will be the approach used in Chapter 9, and the second one is used in the present chapter. Both approaches have their own advantages and drawbacks, and these will be discussed in Chapter 9.

For **K**, we want **KXY** to have the same type as **X**. Suppose **X** has type α and **Y** has type β. Then **KXY** should have type β. This can be achieved by rule (ii) above if **KX** has type $\mathbf{F}\beta\alpha$. And **KX** has this type if we give **K** the type $\mathbf{F}\alpha(\mathbf{F}\beta\alpha)$. This discussion should apply for all α and β, so we must either give $\mathbf{K}$ an infinity of types $\mathbf{F}\alpha(\mathbf{F}\beta\alpha)$, or replace $\mathbf{K}$ by an infinity of restricted combinators $\mathbf{K}_{\alpha\beta}$.

For $\mathbf{S}$ the situation is more complicated. We want $\mathbf{S}XYZ$ to have the same type as $XZ(YZ)$; but first, the fact that $XZ(YZ)$ has a type at all imposes restrictions on the types of X, Y and Z, as follows. Let γ be the type of $(XZ)(YZ)$. Then since types were assumed to be given to composite terms only by rule (ii) above, XZ must have type $\mathbf{F}\beta\gamma$ and YZ must have type β. But for XZ to have type $\mathbf{F}\beta\gamma$, X must have type $\mathbf{F}\alpha(\mathbf{F}\beta\gamma)$, where α is the type of Z. And for YZ to have a type β, Y must have type $\mathbf{F}\alpha\beta$. Summarizing; X, Y, Z must respectively have types of form

$$(1) \qquad \mathbf{F}\alpha(\mathbf{F}\beta\gamma), \qquad \mathbf{F}\alpha\beta, \qquad \alpha .$$

Now we want $\mathbf{S}XYZ$ to have type γ, the same as $XZ(YZ)$. This is achieved by giving $\mathbf{S}$ the type

$$(2) \qquad \mathbf{F}(\mathbf{F}\alpha(\mathbf{F}\beta\gamma))(\mathbf{F}(\mathbf{F}\alpha\beta)(\mathbf{F}\alpha\gamma)) \;;$$

because then by rule (ii) and the type of X in (1), $\mathbf{S}X$ has type

$$\mathbf{F}(\mathbf{F}\alpha\beta)(\mathbf{F}\alpha\gamma) ,$$

and hence by rule (ii) and the type of Y in (1), $\mathbf{S}XY$ has type

$$\mathbf{F}\alpha\gamma ,$$

and finally $\mathbf{S}XYZ$ has type γ as required. Just as with $\mathbf{K}$ and $\mathbf{I}$, we can either give $\mathbf{S}$ every type with form (2), or replace $\mathbf{S}$ by an infinity of restricted combinators $\mathbf{S}_{\alpha\beta\gamma}$, each with a unique type. The former course leads to Chapter 9, and the latter course leads to Definition 8. 2.

Substitution of a term N for a variable x (cf. Definition 2.2) is now only defined when N and x have the same type; in this case [N/x]M will turn out to have the same type as M. Also, if in a term X, a component Y is replaced by a Y' with the same type as Y, the result will be a term with the same type as X.

Definition 8.4 (Typed weak reduction and equality).

Axiom-schemes:

(S) $\mathbf{S}_{\alpha\beta\gamma} X^{\mathbf{F}\alpha(\mathbf{F}\beta\gamma)} Y^{\mathbf{F}\alpha\beta} Z^{\alpha} \rhd X^{\mathbf{F}\alpha(\mathbf{F}\beta\gamma)} Z^{\alpha} (Y^{\mathbf{F}\alpha\beta} Z^{\alpha})$;

(K) $\mathbf{K}_{\alpha\beta} X^{\alpha} Y^{\beta} \rhd X^{\alpha}$;

(I) $\mathbf{I}_{\alpha} X^{\alpha} \rhd X^{\alpha}$;

(ρ) $X \rhd X$.

Definition-rules:

(μ) $X^{\alpha} \rhd Y^{\alpha} \Rightarrow (Z^{\mathbf{F}\alpha\beta} X^{\alpha}) \rhd (Z^{\mathbf{F}\alpha\beta} Y^{\alpha})$;

(ν) $X^{\mathbf{F}\alpha\beta} \rhd Y^{\mathbf{F}\alpha\beta} \Rightarrow (X^{\mathbf{F}\alpha\beta} Z^{\alpha}) \rhd (Y^{\mathbf{F}\alpha\beta} Z^{\alpha})$;

(τ) $X \rhd Y,\ Y \rhd Z \Rightarrow X \rhd Z$.

Equality is defined by the above clauses (with '=' instead of '$\rhd$'), together with the rule

(σ) $X = Y \Rightarrow Y = X$.

Theorem 8.5. If X = Y, then X and Y have the same type.

Proof. By induction on Definition 8.4. For the axiom-schemes the proof is similar to Remark 8.3, and for the deduction-rules it is trivial.

Definition 8.6 (Abstraction). For each variable x with type α, and each term M with type β, we define a term [x]M with type $\mathbf{F}\alpha\beta$, by induction on M as follows.

(i) If $M \equiv x$ (and hence $\alpha \equiv \beta$):

$[x]x \equiv \mathbf{I}_\alpha$. (This has type $\mathbf{F}\alpha\alpha$.)

(ii) If $x \notin M$:

$[x]M \equiv \mathbf{K}_{\beta\alpha}M$. (This has type $\mathbf{F}\alpha\beta$.)

(iii) If $M \equiv Ux$ and $x \notin U$:

$[x]Ux \equiv U$.

(Since x has type α and Ux is given type β by Definition 8.2(ii), U must have type $\mathbf{F}\alpha\beta$.)

(iv) If $M \equiv UV$ and (ii), (iii) do not apply:

$[x]UV \equiv \mathbf{S}_{\alpha\delta\beta}([x]U)([x]V)$

where δ is the type of V. (Since UV is given type β by Definition 8.2(ii), U must have type $\mathbf{F}\delta\beta$; hence by the induction hypothesis, $[x]U$ has type $\mathbf{F}\alpha(\mathbf{F}\delta\beta)$, and $[x]V$ has type $\mathbf{F}\alpha\delta$. Then by the type of $\mathbf{S}_{\alpha\delta\beta}$ and Definition 8.2(ii), $[x]UV$ has type $\mathbf{F}\alpha\beta$.)

Thus abstraction can be defined for all typed M despite the type-restrictions on the basic combinators, and $[x]M$ does have the type we would expect if we considered it as representing a function of x. And by a proof like that of Theorem 2.12, if N has the same type as x, we get

$$([x]M)N \rhd [N/x]M .$$

Also all the other relevant results in Chapter 2 (2.4, 2.8, 2.9, 2.16, 2.17, 2.18, 2.20, 2.21 and their corollaries) hold for this typed system, with hardly any change in their proofs.

In addition we have the following two results, the first very simple, and the second much deeper.

__**Lemma** 8.7.__ __If__ $x_1, \ldots, x_n$ __have types__ $\alpha_1, \ldots, \alpha_n$ __respectively,__ __and__ M __has type__ β, __then__ $[x_1, \ldots, x_n]M$ __has type__

$\mathbf{F}_n\alpha_1 \ldots \alpha_n\beta$.

Proof. By induction on n, using the result in Definition 8.6.

(Notice that $\mathbf{F}_n\alpha_1 \ldots \alpha_n\beta$ can be thought of as representing a set of n-argument functions; if U has type $\mathbf{F}_n\alpha_1 \ldots \alpha_n\beta$ and V_i has type α_i for $i = 1, \ldots, n$, then $UV_1 \ldots V_n$ has type β.)

__**Theorem** 8.8.__ __Every typed term has a normal form with respect to weak reduction.__

Proof. See Sanchis [FDR] Section 4, omitting all mention of **R**. Note that with this omission, the proof becomes finitary.

__**Corollary** 8.8.1.__ __Equality of typed terms is decidable.__

Proof. By Theorem 8.8 and Corollary 2.21.2.

This completes the main properties of the typed system. In Chapter 11 we shall see how it can be extended and applied in a particular context.

Remark 8.9. By analogy with Chapter 3, it is natural to ask how many of the recursive functions can be defined by typed combinators. Suppose we represent the natural numbers by a sequence

$$\mathbf{Z}_n \equiv (\mathbf{SB})^n(\mathbf{KI}) \equiv (\mathbf{S}(\mathbf{S}(\mathbf{KS})\mathbf{K}))^n(\mathbf{KI}) ,$$

where the **S**'s, **K**'s and **I** are chosen so that each $\mathbf{Z}_n$ does have a type. Then Theorem 8.8 shows that only total functions can be defined in the strong sense. But the complete answer to the question is not known precisely (at least, not to the authors). See [CLg. II], §13D3.

For the situation when the numbers are represented by

$$\bar{n} \equiv \bar{\sigma}^n \bar{0}$$

where $\bar{\sigma}$ and $\bar{0}$ are atoms, a partial answer is given in Chapter 11.

Remark 8.10 (Typed λ-terms). Suppose that instead of building up [x]M by combinators we want to postulate λx. M as a primitive term-forming operation, as in Chapter 1. Then we define typed λ-terms inductively as follows.

(i) All the variables v^α and constants c^δ are typed λ-terms, with types α, δ respectively.

(ii) If X has type $\mathbf{F}\alpha\beta$ and Y has type α, then (XY) is a term with type β.

(iii) If Y has type β, then $(\lambda v^\alpha . Y)$ is a term with type $\mathbf{F}\alpha\beta$.

From this definition it can be seen that each typed λ-term has a unique type. Substitution, [N/x]M, is defined as in Definition 1.4, but only when N has the same type as x. Reduction and equality are defined as in Definitions 1.9 and 1.11, but now in the axiom-schemes v must have the same type as y, and N must have the same type as x; also in the deduction-rules X and Z must be restricted so that XZ (or ZX) does have a type. It can easily be proved that equal λ-terms have the same type, and results 1.8, 1.10, 1.12, 1.13 and their corollaries hold for the typed system too. Finally a straightforward adaptation of Sanchis [FDR] section 4 shows that every typed λ-term has a normal form.

In the previously mentioned logical type-theories it is usually λ-terms that are used, not combinatory terms.

Remark 8.11. Strong reduction of typed combinatory terms can be defined by adding to Definition 8.4 the clause

$$X \succ Y \Rightarrow [x].X \succ [x].Y .$$

It can probably be shown (cf. Theorem 9.21) that every typed term has a strong normal form, though the authors have not yet seen the details written out.

9. Combinators with Types: Second Approach

In Remark 8.3 we saw that there are two ways of assigning types to terms: to postulate an infinite number of basic combinators giving each a unique type, or to postulate three basic combinators giving each an infinite number of types. The uniqueness of types in the first approach is a technical advantage, but for attempting to see whether mathematical logic can be based on a generalized concept of function (cf. Remark 2.23), the combinators $\mathbf{I}_{\alpha}$, $\mathbf{K}_{\alpha\beta}$, and $\mathbf{S}_{\alpha\beta\gamma}$ are too restricted. The second approach is more promising for this attempt.

Another question which leads us to the second approach arises naturally out of Remark 8.9; how do we know that in defining

$$\mathbf{Z}_n \equiv (\mathbf{SB})^n(\mathbf{KI}), \qquad \mathbf{B} \equiv \mathbf{S(KS)K},$$

the types of the basic combinators can be chosen so that $\mathbf{Z}_n$ is a typed term? More generally, is there an algorithm for deciding whether a term in the untyped system has a typed analogue or not?

This chapter will describe the second approach to types, and is based on the work of Curry and Seldin. It will be divided into three sections, §A, §B, and §C; §A will set out the basic results (including the above-desired algorithm), and §§B-C will show how Gentzen's techniques can be introduced and deeper results obtained from them. For further reading on this topic, see the sections on functionality in [CLg]; Volume I §§8C-E, 9A-E, 9F2, and Volume II Chapter 14. Some of the basic techniques are explained in more detail for similar systems in [FML] §§5C, 5D, and [CLg. II] §§12C1-4.

Notation and conventions. Terms here will be the combinatory terms of Chapter 2, but now they will include the types. This means assuming that the atomic terms include **F** and the basic types, and defining the composite type $(\mathbf{F}\alpha\beta)$ to be the term $((\mathbf{F}\alpha)\beta)$. (One reason for doing this is that in a logical system we might want to assign types to the types as well as to other terms; another is to make it easier to incorporate our results into the systems of Chapter 10.) As before, lower case Greek letters will denote types.

Although the basic types and **F** are constant atomic terms, they have no reduction properties as do **I**, **K** and **S**. In fact, any atomic constant distinct from **I**, **K**, and **S** will behave like variables with respect to reduction; in most of Chapters 2 to 8, 'variable' could be replaced by 'atomic constant distinct from **I**, **K** and **S**' without any ill effect. Such atomic constants will be called <u>(constant) nonredex atoms</u>. (Cf. [CLg. II] §11F, 'C-indeterminates'.) One important property they have (from Lemma 7.22) is that if the head of X is a nonredex atom, say

$$X \equiv aX_1 \dots X_m ,$$

and $X \succ Y$, then Y has form

$$Y \equiv aY_1 \dots Y_m$$

where $X_i \succ Y_i$ for $i = 1, \dots, m$.

The phrase 'α is assigned to X', or 'α is a type of X', will usually be written

(1) $\vdash \alpha X$;

for the moment this is just notation, but it is motivated by thinking of α as representing a one-argument predicate instead of a set, and reading 'αX' as 'α holds for X' instead of 'X is a member of α'. (The symbol '$\vdash$' represents the informal predicate of assertion, and may be read 'it is true that...' or 'it is a theorem that...'; it is used to convert the term αX into a statement.) We shall sometimes call **X** the subject(-term) and α the predicate(-term).

A basis B will be any set of statements of the form (1).

Finally, for ease of comparison with Curry's work, we shall use the same names as [CLg] for many of the concepts and rules in this chapter and Chapter 10.

9A. BASIC PROPERTIES

Definition 9.1 (Weak assignment of types to combinators). We shall say that α is (weakly) assigned to a combinator **X**, or

(2) $\quad \vdash^{\mathbf{F}} \alpha\mathbf{X}$,

iff the statement $\vdash \alpha\mathbf{X}$ follows from the following axiom-schemes and rule:

Axiom-schemes:

(**FI**) $\quad \vdash \mathbf{F}\alpha\alpha\mathbf{I}$,

(**FK**) $\quad \vdash \mathbf{F}\alpha(\mathbf{F}\beta\alpha)\mathbf{K}$,

(**FS**) $\quad \vdash \mathbf{F}(\mathbf{F}\alpha(\mathbf{F}\beta\gamma))(\mathbf{F}(\mathbf{F}\alpha\beta)(\mathbf{F}\alpha\gamma))\mathbf{S}$,

Rule **F**: $\quad \vdash \mathbf{F}\alpha\beta\mathbf{X}$ and $\vdash \alpha\mathrm{U} \Rightarrow \vdash \beta(\mathbf{X}\mathrm{U})$.

From here on, we shall abbreviate Rule **F** to

$\mathbf{F}\alpha\beta\mathrm{X}, \quad \alpha\mathrm{U} \vdash \beta(\mathrm{XU})$.

In general, $A_1, \ldots, A_m \vdash B$ will be an abbreviation for

$\vdash A_1$ and ... and $\vdash A_m \Rightarrow \vdash B$.

This definition is just a formalization of the first alternative in Remark 8.3, and Rule **F** is the same as Definition 8.2(ii). However, we have not allowed for axioms giving types of atoms other than **I**, **K**, and **S**; e.g., if $\bar{0}$ and $\bar{\sigma}$ are atoms and **N** a basic type, we will want to postulate, among others, the following axioms:

$$\vdash \mathbf{N}\bar{0} \ ,$$
$$\vdash \mathbf{FNN}\bar{\sigma} \ .$$

Such axioms can be introduced as follows.

Definition 9.2 <u>(Weak assignment of types to terms in general)</u>. If B is a basis, we say that

(3) $\quad B \vdash^{\mathbf{F}} \alpha X$

iff the statement $\vdash \alpha X$ is deducible by Rule **F** from the axiom-schemes (**FI**), (**FK**), (**FS**) and the statements in B.

Interpretation. In contrast to Chapter 8, where 'X has type $\mathbf{F}\alpha\beta$' was interpreted as 'X is a function from α to β', we now have to interpret '$\vdash \mathbf{F}\alpha\beta X$' as 'X is a generalized function (cf. Remark 2.23) whose restriction to arguments in α produces values in β', since a term X may now have many types.

Remark. In the $\mathbf{F}_n$ notation introduced in Chapter 8 the axiom-schemes (**FK**) and (**FS**) can be abbreviated as follows:

(**FK**) $\quad \vdash \mathbf{F}_2\alpha\beta\alpha\mathbf{K} \ ,$

(**FS**) $\quad \vdash \mathbf{F}_3(\mathbf{F}_2\alpha\beta\gamma)(\mathbf{F}\alpha\beta)\alpha\gamma\mathbf{S} \ .$

Definition 9.3 <u>(Stratified term)</u>. Let B be a basis in which no variable occurs. A constant term X is said to be <u>stratified relative to</u> B iff

$$B \vdash^{F} \alpha X$$

for some α. If B is empty, then X is said to be stratified. A term X in which the distinct variables which occur are $x_1, \ldots, x_n$ is said to be stratified relative to B iff there exist types $\xi_1, \ldots, \xi_n, \alpha$ such that

$$B, \xi_1 x_1, \ldots, \xi_n x_n \vdash^{F} \alpha X$$

If B is empty, X is stratified.

Example 9.4. The combinator $\mathbf{B} \equiv \mathbf{S}(\mathbf{KS})\mathbf{K}$ is stratified, and for all α, β, and γ,

$$\vdash^{F} \mathbf{F}(\mathbf{F}\beta\gamma)(\mathbf{F}(\mathbf{F}\alpha\beta)(\mathbf{F}\alpha\gamma))\mathbf{B} .$$

Proof. Let α, β, γ be given types and let

$$\alpha_1 \equiv \mathbf{F}(\mathbf{F}\alpha(\mathbf{F}\beta\gamma))(\mathbf{F}(\mathbf{F}\alpha\beta)(\mathbf{F}\alpha\gamma)) ,$$
$$\beta_1 \equiv \mathbf{F}\beta\gamma .$$

Then by axiom-scheme **(FK)** ,

$$\vdash^{F} \mathbf{F}\alpha_1(\mathbf{F}\beta_1\alpha_1)\mathbf{K} .$$

Also, by axiom-scheme **(FS)** ,

$$\vdash^{F} \alpha_1\mathbf{S} .$$

Hence, by Rule **F**,

$$(4) \qquad \vdash^{F} \mathbf{F}\beta_1\alpha_1(\mathbf{KS}) .$$

Now let

$$\alpha_2 \equiv \beta_1 \equiv \mathbf{F}\beta\gamma , \quad \beta_2 \equiv \mathbf{F}\alpha(\mathbf{F}\beta\gamma) , \quad \gamma_2 \equiv \mathbf{F}(\mathbf{F}\alpha\beta)(\mathbf{F}\alpha\gamma) .$$

Then by axiom-scheme **(FS)**,

$$\vdash^{F} \mathbf{F}(\mathbf{F}\alpha_2(\mathbf{F}\beta_2\gamma_2))(\mathbf{F}(\mathbf{F}\alpha_2\beta_2)(\mathbf{F}\alpha_2\gamma_2))\mathbf{S}\ .$$

Also, $\mathbf{F}\alpha_2(\mathbf{F}\beta_2\gamma_2) \equiv \mathbf{F}\beta_1\alpha_1$, so by (4) and Rule **F**

$$(5) \qquad \vdash^{F} \mathbf{F}(\mathbf{F}\alpha_2\beta_2)(\mathbf{F}\alpha_2\gamma_2)(\mathbf{S}(\mathbf{KS}))\ .$$

Finally, define

$$\alpha_3 \equiv \alpha_2\ , \quad \beta_3 \equiv \alpha\ .$$

Then by axiom-scheme **(FK)**,

$$\vdash^{F} \mathbf{F}\alpha_3(\mathbf{F}\beta_3\alpha_3)\mathbf{K}\ .$$

Also, $\beta_2 \equiv \mathbf{F}\alpha(\mathbf{F}\beta\gamma) \equiv \mathbf{F}\alpha\beta_1 \equiv \mathbf{F}\beta_3\alpha_3$, so by Rule **F** and (5),

$$\vdash^{F} \mathbf{F}\alpha_2\gamma_2(\mathbf{S}(\mathbf{KS})\mathbf{K})\ .$$

Using the definitions of α_2 and γ_2, we can easily see that this is the desired result.

The above proof can be written out in tree form by following the construction of **S(KS)K** thus:

$$\dfrac{\dfrac{\vdash \mathbf{F}(\mathbf{F}\beta_1\alpha_1)(\mathbf{F}(\mathbf{F}\alpha_2\beta_2)(\mathbf{F}\alpha_2\gamma_2))\mathbf{S} \qquad \dfrac{\vdash \mathbf{F}\alpha_1(\mathbf{F}\beta_1\alpha_1)\mathbf{K} \qquad \vdash \alpha_1\mathbf{S}}{\vdash \mathbf{F}\beta_1\alpha_1(\mathbf{KS})}}{\vdash \mathbf{F}(\mathbf{F}\alpha_2\beta_2)(\mathbf{F}\alpha_2\gamma_2)(\mathbf{S}(\mathbf{KS}))} \qquad \vdash \mathbf{F}\alpha_2\beta_2\mathbf{K}}{\vdash \mathbf{F}\alpha_2\gamma_2(\mathbf{S}(\mathbf{KS})\mathbf{K})\ .}$$

Each horizontal line here denotes an application of Rule **F**.

Example 9.5. **SII** is not stratified.

Proof. If we could deduce

$$\vdash \eta(\mathbf{SII})$$

for some η, then this must be the conclusion of an inference by Rule **F** whose premises are

$$\vdash \mathbf{F}\xi\eta(\mathbf{SI}) , \qquad \vdash \xi\mathbf{I} ,$$

for some ξ. Since $\vdash \xi\mathbf{I}$ must be an instance of axiom-scheme (**FI**), ξ must have the form $\mathbf{F}\zeta\zeta$, and hence the first of these premises must be

$$\vdash \mathbf{F}(\mathbf{F}\zeta\zeta)\eta(\mathbf{SI}) .$$

Again, this must be the conclusion of an inference whose premises are

$$\vdash \mathbf{F}\varepsilon(\mathbf{F}(\mathbf{F}\zeta\zeta)\eta)\mathbf{S} , \qquad \vdash \varepsilon\mathbf{I} ,$$

and since $\vdash \varepsilon\mathbf{I}$ is an instance of axiom-scheme (**FI**), we must have $\varepsilon \equiv \mathbf{F}\delta\delta$ for some δ. Thus, the first premise must be

$$\vdash \mathbf{F}(\mathbf{F}\delta\delta)(\mathbf{F}(\mathbf{F}\zeta\zeta)\eta)\mathbf{S} .$$

But this must be an instance of (**FS**); hence, for some α, β, γ,

$$\mathbf{F}\delta\delta \equiv \mathbf{F}\alpha(\mathbf{F}\beta\gamma) , \qquad \mathbf{F}\zeta\zeta \equiv \mathbf{F}\alpha\beta , \qquad \eta \equiv \mathbf{F}\alpha\gamma .$$

From these we get

$$\alpha \equiv \delta \equiv \mathbf{F}\beta\gamma, \qquad \alpha \equiv \zeta \equiv \beta ,$$

and hence

$$\mathbf{F}\beta\gamma \equiv \beta ,$$

which is impossible.

Example 9.6. $\mathbf{KK}$ is stratified, and for any α, β, γ,

$$\vdash^{\mathbf{F}} \mathbf{F}\alpha(\mathbf{F}\beta(\mathbf{F}\gamma\beta))(\mathbf{KK}) .$$

Proof. Let $\alpha_1 \equiv \mathbf{F}\beta(\mathbf{F}\gamma\beta)$, $\beta_1 \equiv \alpha$. Then we have

$$\frac{\vdash \mathbf{F}\alpha_1(\mathbf{F}\beta_1\alpha_1)\mathbf{K} \qquad \vdash \alpha_1\mathbf{K}}{\vdash \mathbf{F}\beta_1\alpha_1(\mathbf{KK})} .$$

Example 9.7. The term xx is not stratified.

Proof. A deduction of $\vdash^{\mathbf{F}} \alpha(\mathrm{xx})$ would have to have the form

$$\frac{\vdash \mathbf{F}\beta\alpha\mathrm{x} \qquad \vdash \beta\mathrm{x}}{\vdash \alpha(\mathrm{xx})} .$$

Since Definition 9.3 allows us to assign only one type to a variable, this deduction requires

$$\beta \equiv \mathbf{F}\beta\alpha ,$$

which is impossible.

Remark 9.8. There is an algorithm for deciding for any term $\mathbf{X}$ whether or not $\mathbf{X}$ is stratified. Although this algorithm is simple to operate (cf. Examples 9.5 and 9.7), it is a bit complicated to describe formally. (See Curry [MBF] Theorem 1 or Hindley [PTO] Corollary 1.1.) It depends on the fact that the tree-structure of a deduction

$$\xi_1\mathrm{x}_1, \ldots, \xi_n\mathrm{x}_n \vdash^{\mathbf{F}} \alpha\mathbf{X}$$

is isomorphic to the construction of $\mathbf{X}$; if $\mathbf{X}$ is an atom, then $\vdash \alpha\mathbf{X}$ must be either an axiom-scheme or a statement $\vdash \xi_i\mathrm{x}_i$, whereas if $\mathbf{X} \equiv \mathrm{UV}$ then $\vdash \alpha\mathbf{X}$ must be the conclusion of an inference by Rule $\mathbf{F}$ with premises

$$\vdash \mathbf{F}\beta\alpha\mathrm{U}\ , \qquad \vdash \beta\mathrm{V}$$

for some β.

The algorithm also applies to stratification relative to a basis B under certain conditions; sufficient conditions are set out in the above references.

The parallel between a deduction

$$(6) \qquad \mathrm{B} \vdash^{\mathbf{F}} \alpha\mathbf{X}$$

and the construction of $\mathbf{X}$ means that if a basis B does not contain every atom of $\mathbf{X}$ in its subjects, then we cannot have (6). And if $\mathbf{X}$ is stratified relative to B, every part of $\mathbf{X}$ must also be stratified relative to B.

Exercise 9.9. Prove that **SKSI** is stratified and that all its types have the form

$$\mathbf{F}(\mathbf{F}\beta\gamma)(\mathbf{F}\beta\gamma)\ .$$

Exercise 9.10 (cf. Example 9.4). Prove that every type of $\mathbf{B} \equiv \mathbf{S}(\mathbf{KS})\mathbf{K}$ has the form

$$\mathbf{F}(\mathbf{F}\beta\gamma)(\mathbf{F}(\mathbf{F}\alpha\beta)(\mathbf{F}\alpha\gamma))\ .$$

The following is a list of some useful terms with the most general types obtainable for them.

Term		Type
C	(Ex. 2.6)	$\mathbf{F}(\mathbf{F}\beta(\mathbf{F}\alpha\gamma))(\mathbf{F}\alpha(\mathbf{F}\beta\gamma))$
W	(Ex. 2, Ch. 2)	$\mathbf{F}(\mathbf{F}\alpha(\mathbf{F}\alpha\beta))(\mathbf{F}\alpha\beta)$
$\mathbf{Z}_0$	(Def. 3.1)	$\mathbf{F}\beta(\mathbf{F}\alpha\alpha)$

Term	Type
SB	$\mathbf{F}(\mathbf{F}(\mathbf{F}\beta\gamma)(\mathbf{F}\alpha\beta))(\mathbf{F}(\mathbf{F}\beta\gamma)(\mathbf{F}\alpha\gamma))$
$\mathbf{Z}_1$	$\mathbf{F}(\mathbf{F}\alpha\beta)(\mathbf{F}\alpha\beta)$
$\mathbf{Z}_n$, $n > 1$	$\mathbf{F}(\mathbf{F}\alpha\alpha)(\mathbf{F}\alpha\alpha)$.

Theorem 9.11 (Abstraction and types). *If* x *does not occur in any statement of* B *and if*

$$B, \xi x \vdash^{F} \eta X ,$$

then

$$B \vdash^{F} \mathbf{F}\xi\eta([x]X) .$$

Remark Cf. [CLg. I] §9D, Corollary 1.1. This is Curry's 'stratification theorem', and is related to Rule **F** the way the deduction theorem is related to modus ponens.

Proof. A straightforward induction on X, with cases corresponding to Definition 2.10 (cf. Definition 8.6). The restriction that x not occur in B implies that whenever x occurs in the deduction of $\vdash \eta X$, its type must be ξ.

Corollary 9.11.1 (cf. Lemma 8.7). *If* $x_1, \ldots, x_n$ *are distinct and do not occur in* B, *and if*

$$B, \xi_1 x_1, \ldots, \xi_n x_n \vdash^{F} \eta X ,$$

then

$$B \vdash^{F} \mathbf{F}_n \xi_1 \ldots \xi_n \eta([x_1, \ldots, x_n]X) .$$

Proof. Theorem 9.11 n times.

Exercise. Prove Example 9.4, using the above corollary and the fact that $\mathbf{B} \equiv [x, y, z]x(yz)$.

Theorem 9.12 (*Subject-reduction theorem*; cf. [CLg] §§9C2, 9C6, 14B2). *Let* B *be a basis in which every subject is irreducible relative to* $\triangleright$ *and in which no subject begins with a combinator. Then if*

$$\mathrm{B} \vdash^{\mathbf{F}} \alpha\mathbf{X}, \qquad \mathbf{X} \triangleright \mathbf{X}',$$

then

$$\mathrm{B} \vdash^{\mathbf{F}} \alpha\mathbf{X}'.$$

Proof. We use the fact that weak reduction is a series of contractions by axiom-schemes (**I**), (**K**), (**S**) of Definition 2.3. We can assume that there is only one contraction, since the general result follows by induction. By the discussion in Remark 9.8, we can assume without loss of generality that **X** is the redex and **X'** its contractum.

Case 1. $\mathbf{X} \equiv \mathbf{I}\mathbf{X}'$. By the assumptions about B, the statement $\vdash \alpha(\mathbf{I}\mathbf{X}')$ is not in B. Hence, it must be the conclusion of an inference by Rule **F** whose premises are

$$\vdash \mathbf{F}\beta\alpha\mathbf{I}, \qquad \vdash \beta\mathbf{X}'.$$

Since the first premise is not in B, it must be an instance of axiom-scheme (**FI**), and so $\beta \equiv \alpha$ and the second premise is the desired result.

Case 2. $\mathbf{X} \equiv \mathbf{K}\mathbf{X}'\mathrm{Y}$. By the assumptions about B, $\vdash \alpha\mathbf{X}$ must be the conclusion of a deduction of the following form:

$$\dfrac{\dfrac{\vdash \mathbf{F}\gamma(\mathbf{F}\beta\alpha)\mathbf{K} \qquad\qquad \vdash \gamma\mathbf{X}'}{\mathbf{F}\beta\alpha(\mathbf{K}\mathbf{X}')} \qquad\qquad \vdash \beta\gamma}{\vdash \alpha(\mathbf{K}\mathbf{X}'\mathrm{Y})\,.}$$

Since the top left statement must be an instance of **(FK)**, we have $\gamma \equiv \alpha$ and the top right statement is the desired result.

Case 3. $X \equiv \mathbf{S}UVW$, $X' \equiv UW(VW)$. By an argument like the previous cases, the deduction must be as follows

$$\cfrac{\cfrac{\cfrac{\vdash \mathbf{F}_3(\mathbf{F}_2\gamma\beta\alpha)(\mathbf{F}\gamma\beta)\gamma\alpha\mathbf{S} \qquad \vdash \mathbf{F}_2\gamma\beta\alpha U}{\vdash \mathbf{F}_2(\mathbf{F}\gamma\beta)\gamma\alpha(\mathbf{S}U)} \qquad \vdash \mathbf{F}\gamma\beta V}{\vdash \mathbf{F}\gamma\alpha(\mathbf{S}UV)} \qquad \vdash \gamma W}{\vdash \alpha(\mathbf{S}UVW)\ ,}$$

where we have already used the fact that the top left statement is an instance of **(FS)**. Then, using the types for U, V, and W, we have

$$\cfrac{\cfrac{\vdash \mathbf{F}_2\gamma\beta\alpha U \qquad \vdash \gamma W}{\vdash \mathbf{F}\beta\alpha(UW)} \qquad \cfrac{\vdash \mathbf{F}\gamma\beta V \qquad \vdash \gamma W}{\vdash \beta(VW)}}{\vdash \alpha(UW(VW))\ ,}$$

which is the desired result.

Corollary 9.12.1. <u>The theorem remains true if $\triangleright$ is replaced by $\succ$ throughout.</u>

Proof. See [CLg. I] §9C6.

Remark 9.13. Theorem 9.12 cannot be reversed; it is not, in general, true that if $\vdash \alpha X$ and $X' \triangleright X$, then $\vdash \alpha X'$. For example, take

$X \equiv \mathbf{KI}(\mathbf{SI})$; $\qquad X' \equiv \mathbf{SKSI}$;

then $\vdash^{F} \mathbf{F}\delta\delta X$ holds for all δ, but by Exercise 9.9 $\vdash^{F} \mathbf{F}\delta\delta X'$ holds only for composite δ. An even stronger example is

$$X \equiv \mathbf{\Pi}(\mathbf{\Pi}), \qquad X' \equiv \mathbf{SIII};$$

in this case $\vdash^{F} \mathbf{F}\delta\delta X$ holds for all δ, yet X' is not stratified at all. Thus, in contrast to Chapter 8, the set of types belonging to a term is not invariant with respect to equality. See [CLg. I] §9C3 for special conditions under which Theorem 9.12 can be reversed.

Remark 9.14 (Relation between the two approaches). The stratified combinators here correspond to the typed combinators in Chapter 8. But the most natural correspondence is not one-to-one, because, for example, **K** corresponds to an infinite number of terms $\mathbf{K}_{\alpha\beta}$. However, for each stratified combinator X there is at least one deduction of $\vdash \eta X$ for various η. And to each such deduction corresponds a typed term Y in Chapter 8 as follows:

(i) if $\vdash \eta X$ is an instance of (**FI**), (**FK**), or (**FS**), then Y is the corresponding $\mathbf{I}_{\alpha}$, $\mathbf{K}_{\alpha\beta}$, or $\mathbf{S}_{\alpha\beta\gamma}$;

(ii) if $X \equiv X_1X_2$ and $\vdash \eta X$ is deduced from $\vdash \mathbf{F}\xi\eta X_1$ and $\vdash \xi X_2$, then $Y \equiv Y_1Y_2$ where Y_i corresponds to the given deduction for X_i.

This correspondence, between deductions in Chapter 9 and typed terms in Chapter 8, is one-to-one. (Cf. [CLg. II] §13D5, Theorem 6.)

Remark 9.15. So far, we have not said much about the kinds of statements $\vdash \alpha X$ we want to have in bases. In fact, the only restriction we have so far is that of Theorem 9.12, which is that X be irreducible and not begin with a combinator. Since every term can be uniquely written in the form $X_0X_1 \ldots X_n$ where X_0 is an atom (called the head of the term), this restriction means that the head of X in any statement $\vdash \alpha X$ in the basis must be an atomic term different from **I**, **K**, **S**. Since the natural way to get a type for YZ is from types for Y and Z, this suggests that a basis B should consist of statements $\vdash \alpha X$ only when X is a constant nonredex atom.

Bases are usually defined by specifying one 'axiom-scheme' for each atom X. However, for some purposes we may want to consider bases in which some atoms have more than one scheme. For example, we might want to include in a basis a statement

$$(7) \qquad \vdash \mathbf{F}\alpha\beta\mathbf{I}$$

for particular distinct α and β (cf. [CLg. II] §17C2); this says intuitively that the identity function, when restricted to arguments in α, has values in β, or in other words that α is a subset of β. However, this axiom violates the conditions of Theorem 9.12, and if $\alpha \not\equiv \beta$ we could, from $\vdash \alpha X$, deduce $\vdash \beta(\mathbf{I}X)$ but not necessarily $\vdash \beta X$.

In some situations (cf. Chapter 10), we shall find it useful to be able to deduce $\vdash \beta X$ from $\vdash \alpha X$ if (7) is assumed. To allow this, we can extend Definitions 9.1 and 9.2 by saying that α is <u>strongly assigned</u> to X by the basis B, or

$$(8) \qquad B \vdash^{A} \alpha X ,$$

iff there is a deduction of $\vdash \alpha X$ from B and the axiom-schemes (**FI**), (**FK**), (**FS**), by Rule **F** and

Rule Eq'. $X = Y \quad \& \quad \vdash \xi X \Rightarrow \vdash \xi Y$.

Since it is not decidable whether any two given terms are equal, this new system, unlike those considered so far, is not decidable; hence it would appear that most of the theory of this chapter is useless for it. However, this is not true, for it can be proved that if (8) holds then there is a $Y = X$ such that $B \vdash^{F} \alpha Y$

(i. e., all inferences by Rule Eq' can be pushed down to the end of the deduction and combined into one); see [CLg. I] §9C1. Then, for example, if equality is extensional (but not if it is weak), Theorem 9.11 can be proved for the new system.

9B. A 'NATURAL DEDUCTION' SYSTEM

Now let us look at Theorem 9.11. In any deductive system satisfying this theorem, whatever its axioms and rules (as long as they include rule **F**), the previously given axioms for **I**, **K** and **S** will all be provable. For example, to prove $\vdash \mathbf{F}\alpha\alpha\mathbf{I}$, we start with the trivial deduction

$$\alpha x \vdash \alpha x\ ,$$

and then use the theorem to get a deduction of

$$\vdash \mathbf{F}\alpha\alpha([x]x)\ ,$$

which is the required result. (For the axioms for **S** and **K**, see the proof of Theorem 9.17 later.)

This suggests the following alternative formulation of the assignment of types, based on the same idea as G. Gentzen's 'natural deduction' formulation of logic. (Gentzen [ILD] p. 74.) A deduction will be a tree-form structure as usual, but now there will be a new rule, each of whose applications will 'cancel' one premise above it in the deduction. And when all the premises in a deduction are cancelled, it will be a completed proof.

To formulate this new system, we use the following deduction rules.

Fe: From $\vdash \mathbf{F}\xi\eta U$, $\vdash \xi V$, deduce $\vdash \eta(UV)$.

Fi: From $\vdash \eta Y$, deduce $\vdash \mathbf{F}\xi\eta([x]Y)$, provided that in the deduction of $\vdash \eta Y$, the only premise whose subject contains x is $\vdash \xi x$; and when this rule is used $\vdash \xi x$ must be cancelled wherever it occurs as a premise in the deduction.

Definition 9.16. We say $B \vdash^{T} \alpha X$ iff there is a deduction of $\vdash \alpha X$ (using rules **Fe** and **Fi**), whose uncancelled premises are all in B. If there is a deduction of $\vdash \alpha X$ whose premises are all cancelled, we say $\vdash^{T} \alpha X$.

The two rules in Definition 9.16 can be written out in the usual tree-form as follows, using brackets to indicate the cancelled premises above $\vdash \eta Y$ in **Fi**:

$$\textbf{Fe:}\quad \frac{\vdash \mathbf{F}\xi\eta U \qquad \vdash \xi V}{\vdash \eta(UV)} . \qquad\qquad \textbf{Fi:}\quad \frac{\begin{matrix}[\vdash \xi x]\\ \vdash \eta Y\end{matrix}}{\vdash \mathbf{F}\xi\eta([x]Y)} .$$

Notice that rule **Fe** eliminates an **F**, and rule **Fi** introduces one (into the conclusion).

Theorem 9.17. For any B, α and X,

$$B \vdash^{T} \alpha X \iff B \vdash^{F} \alpha X .$$

Proof. From left to right, the theorem follows by Theorem 9.11.

For the proof from right to left, it is sufficient to prove each axiom of the schemes **(FI)**, **(FK)**, **(FS)** in the system of Definition 9.16. We have already done this informally for **(FI)**; the formal proof can be written out in tree form as follows:

$$\frac{\overset{1}{\vdash \alpha x}}{\vdash \mathbf{F}\alpha\alpha\mathbf{I}}\ \textbf{Fi - 1}$$

where '**Fi** - 1' indicates that the conclusion follows by **Fi** where premise 1 is cancelled.

For (**FS**), we use the fact that $[z]xz(yz) \equiv \mathbf{S}xy$, and we then proceed as follows:

$$\dfrac{\dfrac{\dfrac{\dfrac{\dfrac{\overset{z}{\vdash \mathbf{F}\alpha(\mathbf{F}\beta\gamma)x} \qquad \overset{y}{\vdash \alpha z}}{\vdash \mathbf{F}\beta\gamma(xz)}\,\mathbf{Fe} \qquad \dfrac{\overset{z}{\vdash \mathbf{F}\alpha\beta y} \qquad \overset{y}{\vdash \alpha z}}{\vdash \beta(yz)}\,\mathbf{Fe}}{\vdash \gamma(xz(yz))}}{\vdash \mathbf{F}\alpha\gamma(\mathbf{S}xy)}\,\mathbf{Fi}-1}{\vdash \mathbf{F}(\mathbf{F}\alpha\beta)(\mathbf{F}\alpha\gamma)(\mathbf{S}x)}\,\mathbf{Fi}-2}{\vdash \mathbf{F}_3(\mathbf{F}\alpha(\mathbf{F}\beta\gamma))(\mathbf{F}\alpha\beta)\alpha\gamma\mathbf{S}}\,\mathbf{Fi}-3$$

For (**FK**), we use the fact that $[y]x \equiv \mathbf{K}x$, and we then proceed as follows:

$$\dfrac{\dfrac{\overset{y}{\vdash \alpha x}}{\vdash \mathbf{F}\beta\alpha(\mathbf{K}x)}\,\mathbf{Fi}-\ \{\text{cancelling the non-existent premise } \vdash \beta y\}}{\vdash \mathbf{F}\alpha(\mathbf{F}\beta\alpha)\mathbf{K}}\,\mathbf{Fi}-1 \quad .$$

(Cancelling a non-existent premise is quite in order; rule **Fi** did not say that x must actually occur in a premise.)

Remark. We can get a natural deduction formulation for the system introduced in Remark 9.15 simply by adding Rule Eq' to Definition 9.16.

9C. A GENTZEN L-SYSTEM

Gentzen's natural deduction formulation of logic served as a half-way stage between the usual Hilbert-style formulation and his calculi LJ and LK (cf. his [ILD] §III), for which he obtained an important theorem on the nature of proofs (the cut-elimination theorem) from which several other useful properties, e.g. consistency, follow immediately.

We shall now consider an L-formulation of the assignment of types. Before beginning this, notice that if $B \vdash^{F} \alpha X$ holds, then only a finite number of members of B can occur in the deduction of $\vdash \alpha X$. Hence from here on we shall restrict ourselves to finite sets B. Also, to help in the proof of the elimination theorem, we shall use sequences instead of sets.

Notation. 'M' will denote an arbitrary (perhaps empty) finite sequence of terms of form αX, and the concatenation of sequences M_1, M_2, M_3 (for example) will be denoted simply by

'M_1, M_2, M_3' .

No distinction will be made between a one-member sequence and its one member. Deduction-rules will be written with premises above a horizontal line, and conclusion below. The names of most rules will be the same here as in [CLg. II] Chapters 12 and 14, to aid comparison.

The first step towards an L-system is to notice that Definition 9.16 can be rewritten as a direct inductive definition of $\vdash^{T}$, as follows:

Axiom-scheme: $\xi X \vdash^{T} \xi X$.

Deduction-rules:

(i) $$\frac{B \vdash^{T} \mathbf{F}\xi\eta U \qquad B \vdash^{T} \xi V}{B \vdash^{T} \eta(UV)}\ .$$

(ii) $$\frac{B, \xi x \vdash^{T} \eta Y}{B \vdash^{T} \mathbf{F}\xi\eta([x]Y)}\ , \qquad \text{if } x \text{ does not occur in } B.$$

But these two rules are not enough; we need also

(iii) $$\frac{B \vdash^{T} \xi X}{B' \vdash^{T} \xi X}, \qquad \text{if } B' \supseteq B;$$

(iv) $$\frac{B \vdash^{T} \xi X \qquad B,\ \xi X \vdash^{T} \eta Y}{B \vdash^{T} \eta Y}.$$

This last is called Rule cut, and expresses a sort of transitivity of the relation $\vdash^{T}$. The purpose of the L-formulation is to find new versions of (i) and (ii) which have a fairly simple structure (for example, with every type that appears in the premises also appearing in the conclusion), and which make the cut-rule redundant.

We now define the L-formulation.

Definition 9.18. The relation $\Vdash$ between sequences and terms is defined inductively as follows.

Axiom-scheme: $\xi X \Vdash \xi X$.

Deduction-rules:

*C: $$\frac{M \Vdash \xi X}{M' \Vdash \xi X}, \qquad \text{if } M' \text{ is any permutation of } M.$$

*W: $$\frac{M,\ \eta Y,\ \eta Y \Vdash \xi X}{M,\ \eta Y \Vdash \xi X}.$$

*K: $$\frac{M \Vdash \xi X}{M,\ \eta Y \Vdash \xi X}.$$

F'*: $$\frac{M,\ \xi x \Vdash \eta Y}{M \Vdash \mathbf{F}\xi\eta([x]Y)}, \qquad \text{if } x \text{ does not occur in } M.$$

$$*\mathbf{F}:\quad \frac{M \Vdash \xi V \qquad M,\ \eta(UV) \Vdash \zeta Y}{M,\ \mathbf{F}\xi\eta U \Vdash \zeta Y\ .}$$

$$*\mathbf{F}':\quad \frac{M \Vdash \xi V \qquad M,\ \eta([V/x]X) \Vdash \zeta Y \qquad M,\ \eta(([x]X)V) \Vdash \eta([V/x]X))}{M,\ \mathbf{F}\xi\eta([x]X) \Vdash \zeta Y\ .}$$

The rules *C, *W, *K are called structural rules; the first two ensure that if M_1, M_2 are two sequences formed from the same set of terms, then

$$M_1 \Vdash \xi X \iff M_2 \Vdash \xi X\ .$$

Rule *K corresponds to the earlier rule (iii), and Fule **F'*** is a re-statement of (ii). Rule ***F** seems to express the same property as (i), but with the technical advantage that every type appearing in the premises also appears in the conclusion. Unfortunately ***F** is not strong enough to make $\Vdash$ equivalent to $\vdash^{T}$; it turns out that we need ***F'** as well. (In the special case $X \equiv Ux$ with $x \notin U$, ***F'** is the same as ***F**, so if it were not for the need to motivate ***F'**, ***F** could well be omitted.) The third premise in ***F'** is essentially an assumption that reduction preserves types.

Rules ***F**, ***F'**, **F'*** are called operational rules.

The next theorem shows that the cut rule does not need to be added to the above list.

Theorem 9.19 (Elimination theorem, or ET). If $M \Vdash \xi X$ and $M,\ \xi X \Vdash \eta Y$, then

$$M \Vdash \eta Y\ .$$

Summary of proof. By induction on the number of occurrences of **F** in ξ. Since the present type-theory is not the main one developed by Curry in [CLg. II], but a modification of it, its ET proof is not given in full in the literature. However, a similar proof occurs in [CLg. II] §12C1-4 for a non-type-theoretic system; and in §14E1-2 this proof is modified to apply to type-theory with equality-rules included, while §14E6 gives the modifications necessary for the present system. The main step in the proof is as follows.

By a secondary induction very much like [CLg. II] §12C4 or §14E2, it can be shown that it is enough to consider the case in which the statements $M, \xi X \Vdash \eta Y$ and $M \Vdash \xi X$ are conclusions of inferences by $*\mathbf{F}'$ and $\mathbf{F}'*$ respectively, and where ξX in each is the new constituent introduced into the conclusion by the rule. Then the last step in the deduction of $M, \xi X \Vdash \eta Y$ has the form

$$(9) \qquad \frac{M \Vdash \alpha V \qquad M, \beta([V/x]Z) \Vdash \eta Y \qquad M, \beta(([x]Z)V) \Vdash \beta([V/x]Z)}{M, \mathbf{F}\alpha\beta([x]Z) \Vdash \eta Y}$$

where $\xi \equiv \mathbf{F}\alpha\beta$ and $X \equiv [x]Z$. Also, the last step in the deduction of $M \Vdash \xi X$ has the form

$$(10) \qquad \frac{M, \alpha x \Vdash \beta Z}{M \Vdash \mathbf{F}\alpha\beta([x]Z)}, \quad \text{where } x \notin M, \mathbf{F}\alpha\beta .$$

Hence, by substituting V for x in the proof of the premise of (10), we get

$$(11) \qquad M, \alpha V \Vdash \beta([V/x]Z) .$$

(Incidentally, when V is substituted into an application of rule $\mathbf{F}'*$ in this proof, we must make sure that V does not contain the variable x in the premise of that application. This is done by changing that variable to a new one before substituting V; since x does not occur in the conclusion of $\mathbf{F}'*$, this change will not affect the conclusion. Cf. Curry [FML] §7B Theorem 1.)

Now by (11), the second premise of (9), and the main induction-hypothesis (since β has fewer occurrences of $\mathbf{F}$ than ξ), we get

$$\mathrm{M},\ \alpha\mathrm{V} \Vdash \eta\mathrm{Y}\,.$$

We can then get $\mathrm{M} \Vdash \eta\mathrm{Y}$ by this, the first premise in (9), and another application of the main induction-hypothesis.

Theorem 9.20. *For each finite set* B *let* M *be any arrangement of the members of* B *in a sequence. Then*

$$\mathrm{M} \Vdash \xi\mathbf{X} \Longleftrightarrow \mathrm{B} \vdash^{\mathrm{T}} \xi\mathbf{X}\,.$$

Proof from left to right. The axiom-scheme in Definition 9.18 and the rules $*\mathrm{C}$, $*\mathrm{K}$, $*\mathrm{W}$, $\mathbf{F}'*$ are all true for $\vdash^{\mathrm{T}}$, as we have seen earlier. For $*\mathbf{F}'$, suppose we have

(12) $\quad \mathrm{B} \vdash^{\mathrm{T}} \xi\mathrm{V}$,

(13) $\quad \mathrm{B},\ \eta([\mathrm{V}/x]\mathbf{X}) \vdash^{\mathrm{T}} \zeta\mathrm{Y}$,

(14) $\quad \mathrm{B},\ \eta(([x]\mathbf{X})\mathrm{V}) \vdash^{\mathrm{T}} \eta([\mathrm{V}/x]\mathbf{X})$.

Then we prove

$$\mathrm{B},\ \mathbf{F}\xi\eta([x]\mathbf{X}) \vdash^{\mathrm{T}} \zeta\mathrm{Y}$$

as follows:

$$\dfrac{\mathbf{F}\xi\eta([x]\mathbf{X}) \qquad \dfrac{\mathrm{B}}{\xi\mathrm{V}}\ \text{by (12)}}{\dfrac{\eta(([x]\mathbf{X})\mathrm{V})}{\dfrac{\eta([\mathrm{V}/x]\mathbf{X})}{\zeta\mathrm{Y}\,.}\ \text{by (13)}}\ \text{by (14)}}\ \mathbf{Fe}$$

Proof from right to left. If ξX is in B, then

$$M \Vdash \xi X$$

follows from the axiom $\xi X \Vdash \xi X$ by rules *K and *C.

Rule **Fi** in Definition 9.16 is valid for $\Vdash$ since it is just **F**'*. To show that rule **Fe** is valid for $\Vdash$, suppose that

$$M_1 \Vdash \mathbf{F}\xi\eta U\,, \qquad M_2 \Vdash \xi V\,;$$

then for $M \equiv M_1, M_2$, we have by *K and *C,

$$(15) \qquad M \Vdash \mathbf{F}\xi\eta U\,, \qquad M \Vdash \xi V\,.$$

Now, we can prove the following:

$$\dfrac{\dfrac{\xi V \Vdash \xi V}{M, \xi V \Vdash \xi V}\text{ by *K, *C} \qquad \dfrac{\eta(UV) \Vdash \eta(UV)}{M, \xi V, \eta(UV) \Vdash \eta(UV)}\text{ by *K, *C}}{M, \xi V, \mathbf{F}\xi\eta U \Vdash \eta(UV)\,.}\text{ by *}\mathbf{F}$$

Then $M \Vdash \eta(UV)$ follows by (15) and **ET** twice.

<u>**Corollary 9.20.1.**</u> <u>If</u> M <u>is any sequence of all the members of</u> B, <u>then</u>

$$B \vdash^{F} \xi X \iff M \Vdash \xi X\,.$$

Proof. By Theorems 9.17 and 9.20.

<u>**Theorem 9.21.**</u> <u>Every stratified term has a strong normal form.</u> <u>More generally, let</u> B <u>be a set of terms</u> $\eta_1 U_1, \eta_2 U_2, \ldots,$ <u>and let every</u> U_i <u>have a strong normal form,</u> <u>and no</u> U_i <u>whose</u> η_i <u>is composite be headed by</u> **S**, **K** <u>or</u> **I**; <u>then if</u> $B \vdash^{F} \xi X$, X <u>must have a strong normal form.</u>

Proof. By Corollary 9.20.1, it is enough to show that if a sequence M satisfies the above conditions, and

$$(16) \qquad \mathbf{M} \Vdash \xi\mathbf{X},$$

then **X** has a normal form. This will be done by induction on the proof of (16).

For the basic step, (16) is an axiom and then M is just ξX; hence **X** has a normal form by the assumptions.

For the induction step, suppose first that (16) is the conclusion of an inference by *C, *K or *W; then **X** has a normal form by the induction-hypothesis.

Next, suppose that the proof of (16) ends with

$$\frac{\mathbf{M},\ \alpha x \Vdash \beta\gamma}{\mathbf{M} \Vdash \mathbf{F}\alpha\beta([x]\mathbf{Y})} \quad \text{by } \mathbf{F}'^{*} \quad \xi \equiv \mathbf{F}\alpha\beta, \quad \mathbf{X} \equiv [x]\mathbf{Y}.$$

Then the sequence M, αx satisfies the given conditions, so by the induction-hypothesis, **Y** has a normal form, **Y***. Hence by rule (ξ) in Definition 7.1,

$$\mathbf{X} \equiv [x]\mathbf{Y} \succ [x]\mathbf{Y}^{*}.$$

Since [x]**Y*** is in normal form (Definition 7.18), it is the required normal form of **X**.

Finally, suppose that the last step in the proof of (16) has form

$$\frac{\mathbf{M}_1 \Vdash \alpha\mathbf{V} \qquad \mathbf{M}_1, \beta([\mathbf{V}/x]\mathbf{Z}) \Vdash \xi\mathbf{X} \qquad \mathbf{M}_1, \beta(([x]\mathbf{Z})\mathbf{V}) \Vdash \beta([\mathbf{V}/x]\mathbf{Z})}{\mathbf{M}_1, \mathbf{F}\alpha\beta([x]\mathbf{Z}) \Vdash \xi\mathbf{X}}$$

where M is the sequence $\mathbf{M}_1$, **F**αβ([x]**Z**). By the given condition on M, the term [x]**Z** cannot be headed by **S**, **K** or **I**. Hence (cf. Definition 2.10), we must have

$$Z \equiv Ux \qquad (x \notin U), \qquad [x]Z \equiv U \equiv aU_1 \ldots U_n ,$$

where the atom a is not **S**, **K** or **I**. Also by the assumptions, U must have a strong normal form U*, and by Lemma 7.22 we have

$$U^* \equiv aU_1^* \ldots U_n^* .$$

Now by the induction-hypothesis and the first premise, V has a normal form V*. Then

$$[V/x]Z \equiv UV \succ aU_1^* \ldots U_n^* V^* ,$$

and the latter is in normal form by Definition 7.18. Hence the sequence

$$M_1, \ \beta([V/x]Z)$$

satisfies the given assumptions, so by the induction hypothesis and the second premise above, X has a normal form.

Remark 9.22. If equality is extensional, then it is possible to get an L-system satisfying ET that is equivalent to the system of Remark 9.15 with the equality-rule. Indeed, we remarked at the end of §B on how to get the natural deduction system. The neatest method of getting the L-formulation seems to be to adjoin to Definition 9.18 the following two 'expansion-rules':

Exp $\qquad \dfrac{M, \xi Y \Vdash \eta Z}{M, \xi X \Vdash \eta Z} \ (\text{if } X \succ Y), \qquad \dfrac{M \Vdash \xi Y}{M \Vdash \xi X} \ (\text{if } X \succ Y),$

and to replace Rule **F**'* by

F* $\qquad \dfrac{M, \xi x \Vdash \eta(Xx)}{M \Vdash \mathbf{F}\xi\eta X} \ (\text{if } x \notin M, X) ,$

and to omit Rule ***F'** (but keep ***F**).

These modifications are enough to give a proof that if

$$\eta_1 Y_1, \ldots, \eta_n Y_n \Vdash \xi Y$$

and if $Y = X$ and $Y_i = X_i$ for $i = 1, \ldots, m$, then

$$\eta_1 X_1, \ldots, \eta_n X_n \Vdash \xi X .$$

(Cf. [CLg.II] §12C, Theorem 7.) The elimination theorem can also be proved. (Cf. [CLg. II] §14E4.)

Remark 9. 23. If we wish to assign types to λ-terms instead of combinatory terms, the most natural system to use is the natural deduction one, with $\lambda x.Y$ instead of $[x]Y$ in Definition 9. 16. Then deductions parallel constructions as in Remark 9. 8, and analogues of the subject-reduction theorem (9. 12) and Remark 9. 14 hold for λ-terms. At present, no corresponding L-system for λ-terms is known; but if we add Rule Eq' as in Remark 9. 15, then an equivalent L-system can be formulated (and is obtained from the one in Remark 9. 22 above by replacing $[x]Y$ by $\lambda x. Y$), and the elimination theorem holds (cf. [CLg. II] §§14D-E). In this case, it is no longer necessary that equality be extensional.

10. Logic Based on Combinators

In this chapter, we shall consider the use to which we can put combinators in setting up ordinary systems of formal logic. It turns out that it is sufficient for this purpose to use either combinatory terms or λ-terms, and that there is no advantage in using one rather than the other. Hence, we shall revise our notation as follows, so that it will refer to both systems simultaneously.

Notation	Meaning for λ-terms	For combinatory terms
$\lambda x.\,X$	$\lambda x.\,X$	$[x]X$
$X \equiv Y$	X congruent to Y	$X \equiv Y$
x occurs in Y	x free in Y	x occurs in Y
$X = Y$	$X =_e Y$ (Def. 5.2)	$X =_e Y$
$X \succ Y$	$X \succ_\lambda Y$ (Def. 7.3)	$X \succ Y$ (Def. 7.1)
I, **K**, **S**	$\mathbf{I}_\lambda$, $\mathbf{K}_\lambda$, $\mathbf{S}_\lambda$ (Def. 6.1)	**I**, **K**, **S**
B, **C**, **W**	$\mathbf{B}_\lambda$, $\mathbf{C}_\lambda$, $\mathbf{W}_\lambda$	**B**, **C**, **W**

(The last three combinators are defined in Example 2.5, 2.6 and Exercise 2 Chapter 2.)

In addition to statements involving the predicates $=$ and $\succ$, we will want to consider statements of the form

$$\vdash X,$$

where X is a term. These will be called assertions and they are the statements we shall be most interested in. They will be connected with equality by the following rule:

Rule Eq. $X = Y \;\;\&\;\; \vdash X \;\Rightarrow\; \vdash Y.$

(Compare Rule Eq' in Remark 9.15.) As in Chapter 9, we shall abbreviate

$$\vdash A_1 \quad \& \quad \ldots \quad \& \quad \vdash A_n \Rightarrow \vdash X$$

as

$$A_1, \ldots, A_n \vdash X.$$

Now the naive way to go about setting up a system of logic based on combinatory or λ-terms is just to introduce terms representing the usual connectives and quantifiers and postulate their ordinary rules. Unfortunately, this fails to work. For suppose that to each X and Y corresponds a term $X \supset Y$ which satisfies the following rule and axiom-scheme:

(1) $X \supset Y,\ X \vdash Y$

(2) $\vdash (X \supset (X \supset Y)) \supset (X \supset Y)$.

Then, if we have Rule **Eq**, we can prove the system inconsistent in the sense that $\vdash Y$ holds for every term Y. We prove this as follows: let Y be any term, and define

$$Z \equiv \lambda z.(zz \supset (zz \supset Y)), \qquad (z \notin Y)$$

$$X \equiv ZZ .$$

Then

(3) $X = X \supset (X \supset Y)$

(4) $\quad = (X \supset (X \supset Y)) \supset (X \supset Y)$.

Hence:

	$\vdash (X \supset (X \supset Y)) \supset (X \supset Y)$	by (2);
(5)	$\vdash X \supset (X \supset Y)$	by (4), Rule **Eq**;
(6)	$\vdash X$	by (3), Rule **Eq**;
(7)	$\vdash X \supset Y$	by (5), (6), Rule (1);
	$\vdash Y$	by (6), (7), Rule (1).

Since Y is an arbitrary term, we have the inconsistency of the system. (This argument is known as Curry's Paradox.)

Now, Rule (1) is just the rule modus ponens, which we surely need in any system of logic. Furthermore, the scheme (2) can easily be shown to hold if we have the deduction theorem:

$$A_1, \ldots, A_n, X \vdash Y \quad \Rightarrow \quad A_1, \ldots, A_n \vdash X \supset Y,$$

and the deduction theorem in some form or other is a property of any reasonable implication. So what this paradox shows is that the deduction theorem in its full generality (i.e. for any terms X and Y) is incompatible with the property called combinatory completeness, i.e. that $\lambda x.X$ is a term for every term X. (It does not help to restrict combinatory completeness as Church did by requiring that x occur (free) in X in order for $\lambda x.X$ to be a term, since this paradox still works in that case.)

It thus appears that to get a consistent system of logic based on combinatory terms or λ-terms, we must either restrict the terms X for which $\lambda x.X$ is a term, or else restrict the terms for which the deduction theorem holds. One possible restriction on $\lambda x.X$ is that x have at most one (free) occurrence in X; a system of this kind was explored a little by Fitch in his [SFL], and was found to be insufficient for much of mathematics. In this chapter we shall leave abstraction unrestricted, and formulate restrictions on the deduction theorem.

An important restriction we shall not have the space to consider here is to insist that terms in the deduction theorem must be stratified (cf. Chapter 9). This leads to type-theories, which were first introduced by Church in his [FST] (based on λ-terms), and have been studied by many people since then. For further reading in this topic, see Henkin [CTT], [TPT], Andrews [TTT], Schütte [SSP], [STT], Sanchis [TCL], Prawitz [HHO], Takahashi [PCE], and [CLg. II] Chapter 17. All but [CLg. II] and Sanchis' paper are based on the Chapter 8 approach to types, but Curry and Sanchis showed that the Chapter 9 approach would do also.

Other combinator-based systems of logic have been developed by F. B. Fitch; for information about these, see his [SCD], [CCM] and the references given therein, also Orgass and Fitch [TCM].

10A. LOGICAL CONSTANTS

Before we discuss what form of the deduction theorem to adopt, we must discuss the terms to be introduced to represent the logical connectives and quantifiers.

It turns out that the connectives and quantifiers which are a cause for concern are implication and universal quantification. Therefore, it is natural to begin by postulating a constant non-redex atom **P** to stand for implication, and to take $X \supset Y$ to be an abbreviation for **PXY**. With **P**, we postulate

Rule **P**: $\mathbf{PXY}, \mathbf{X} \vdash \mathrm{Y}$.

Similarly, for universal quantification, we can postulate a non-redex atom Π, together with

Rule Π: $\Pi\mathbf{X} \vdash \mathbf{XY}$

for all terms **X** and Y. Then if we make the abbreviation

$$(\forall x)Z \equiv \Pi(\lambda x.\ Z) ,$$

rules Π and Eq give

$$\begin{aligned}(\forall x)Z &\vdash (\lambda x.\ Z)Y && \text{by rule } \Pi ,\\ &\vdash [Y/x]Z && \text{by rule Eq.}\end{aligned}$$

Now in most contexts where predicate calculus is used, we do not quantify over absolutely everything, but instead over restricted ranges, using formulas of the form

$$(\forall x \in X)(Yx) .$$

If, considering X as a predicate rather than a set, we replace formulas of form '$x \in X$' by 'Xx', we can use '$\supset$' to rewrite the above formula as

$$(\forall x)\ (Xx \supset Yx) .$$

In a notation sometimes used in predicate logic, this can be written as

$$Xx \supset_x Yx .$$

To express this formally as a term, we define

$$\Xi \equiv \lambda xy.\, \Pi(\lambda z.\, \mathbf{P}(xz)(yz)) .$$

Then if $z \notin XY$, we have

$$\begin{aligned} \Xi XY &= \Pi(\lambda z.\, \mathbf{P}(Xz)(Yz)) \\ &= (\forall z)(Xz \supset Yz) . \end{aligned}$$

Then we can define, for all X, Y, Z,

$$X \supset_z Y \equiv \Xi(\lambda z.\, X)(\lambda z.\, Y) .$$

The following rule for Ξ can be deduced from rules Π and $\mathbf{P}$:

<u>Rule</u> Ξ: $\qquad \Xi XY,\ XU \vdash YU .$

This may also be stated as

$$(Xz \supset_z Yz),\ XU \vdash YU \qquad (z \notin X) ,$$

or equivalently, putting $X \equiv \lambda z.\, X'$ and $Y \equiv \lambda z.\, Y'$ and using Rule Eq,

$$(8) \qquad (X' \supset_z Y'),\quad [U/z]X' \vdash [U/z]Y' .$$

Now instead of defining Ξ in terms of Π and $\mathbf{P}$, we can take Ξ as a constant nonredex atom and postulate Rule Ξ for it. Then if we define

$$(9) \qquad \mathbf{P} \equiv \lambda xy.\, \Xi(\mathbf{K}x)(\mathbf{K}y) ,$$

we can deduce Rule $\mathbf{P}$ as follows (where U is any term):

$$\dfrac{\dfrac{\mathbf{P}XY}{\Xi(\mathbf{K}X)(\mathbf{K}Y)}\,\text{Eq} \qquad \dfrac{X}{\mathbf{K}XU}\,\text{Eq}}{\dfrac{\mathbf{K}YU}{Y\,.}\,\text{Eq}}\,\Xi$$

Suppose there is also a universal category, i. e. a term $\mathbf{E}$ such that we have, for all X (including variables),

$$(10) \qquad \vdash \mathbf{E}X .$$

Then we can define Π by

$$(11) \qquad \Pi \equiv \Xi\mathbf{E} ,$$

and Rule Π will hold as follows:

$$\dfrac{\dfrac{\Pi X}{\Xi\mathbf{E}X}\,\text{Def.} \qquad \overset{(10)}{\mathbf{E}Y}}{XY\quad .}\,\Xi$$

So far, then, we have two kinds of systems. In one kind, the nonredex atoms are Π and **P**. We will call a system of this type a system of universal generality (Curry's systems $\mathscr{F}_3$). The second kind of system has Ξ and **E** as the nonredex atoms; we will call this kind of system a system of restricted generality (Curry's systems $\mathscr{F}_2$). The earliest systems studied, by Kleene, Rosser and Curry before the war, were based on Π and **P**. When the more general of these led to inconsistencies, Curry turned to Ξ as more likely to prove manageable. But it has recently been found that, given certain natural axioms, both alternatives lead to systems of essentially the same strength. However, since the more recent published work has mostly been in terms of Ξ, it will be assumed here that Ξ and **E** are the atoms, with Rules Ξ, Eq and the axioms (10) postulated for them. (Though we shall look at the other kind of system in Section D.)

Besides Ξ, **E**, Π and **P**, there are two other constants worth mentioning here. The first is **F**, from Chapter 9. In terms of Ξ, we can define

$$(12) \qquad \mathbf{F} \equiv \lambda xyz.\, \Xi x(\mathbf{B}yz)\ .$$

Then

$$\begin{aligned}\mathbf{F}XYU &= \Xi X(\mathbf{B}YU)\\ &= Xv \supset_v Y(Uv)\ ,\end{aligned}$$

and we can deduce Rule **F** as follows:

$$\dfrac{\dfrac{\mathbf{F}XYU}{\Xi X(\mathbf{B}YU)}\,\mathrm{Eq} \qquad XV}{\dfrac{\mathbf{B}YUV}{Y(UV)}\,\mathrm{Eq}}\,\Xi\ .$$

To get the axiom-schemes (**FI**), (**FK**) and (**FS**) of Chapter 9, we would of course have to postulate some axiom-schemes for Ξ;

ones which are suitable will be given in Section B.

It is interesting to note that given **F** and Rule **F** we can define a Ξ satisfying Rule Ξ; in fact, we can define at least two, namely

$$\Xi' \equiv \lambda xy.\,\mathbf{F}xy\mathbf{I}\ ,$$
$$\Xi'' \equiv \lambda x.\,\mathbf{F}x\mathbf{I}\ .$$

(The proof that these satisfy Rule Ξ is left as an exercise.) This suggests that **F** could be taken as basic in a system of logic, and in fact this is explored in [CLg. I] Chapter 9 and [CLg. II] Chapter 14 (systems of <u>functionality</u>, or $\mathscr{F}_1$). It was also begun in Chapter 9 of these notes; however, there we did not consider Rule Eq, but, rather the more restricted Rule Eq', and it turns out that for technical reasons the use of the full Rule Eq causes difficulties in a theory of functionality. Hence, in what follows, we shall be considering only systems of universal and restricted generality.

The second interesting extra constant is a term **Q** with axioms and rules giving the statement

$$\vdash \mathbf{Q}\mathrm{XY}$$

some of the properties of the statement $\mathrm{X} = \mathrm{Y}$. The simplest axiom-scheme to postulate is

$$(\rho) \qquad \vdash \mathbf{Q}\mathrm{XX}\ .$$

From this, Rule Eq gives

$$(13) \qquad \mathrm{X} = \mathrm{Y} \quad \Rightarrow \quad \vdash \mathbf{Q}\mathrm{XY}\ .$$

Also, given (ρ), we can define a term E satisfying (10) by

$$\mathbf{E} \equiv \lambda x.\,\mathbf{Q}xx\ .$$

We could also postulate the following rule:

Rule **Q**: $\quad \mathbf{Q}XY,\ ZX \vdash ZY$.

From Rule **Q** and (ρ) with Rule Eq, we could then prove the following (see [CLg. I] §7C2):

(σ) $\quad \mathbf{Q}XY \vdash \mathbf{Q}YX$,

(τ) $\quad \mathbf{Q}XY,\ \mathbf{Q}YZ \vdash \mathbf{Q}XZ$,

(μ) $\quad \mathbf{Q}XY \vdash \mathbf{Q}(ZX)(ZY)$,

(ν) $\quad \mathbf{Q}XY \vdash \mathbf{Q}(XZ)(YZ)$.

Hence, from (ρ), Rule **Q** and Rule Eq, we could derive all the axioms and rules of weak equality with $\vdash \mathbf{Q}XY$ in place of $X = Y$. (Rule **Q** corresponds to what is known as Leibnitz' principle: if $X = Y$, then every predicate Z holding true for X also holds true for Y.)

If it should happen that the converse of (13) holds, then we have a consistency result, since it follows that we do not have $\vdash \mathbf{QSK}$. This kind of consistency, which is fairly strong, is called **Q**-consistency.

10B. THE DEDUCTION THEOREM

We now turn our attention to finding, in a system whose atoms are Ξ and **E**, axioms which allow us to prove a suitably restricted form of the deduction theorem. The first step is to describe the class of terms for which we expect the deduction theorem to hold; i.e. the terms which will be interpreted as propositions.

Clearly, to do this we will want to describe the terms we consider as propositional functions, and those we consider as possible subjects of these functions. The terms to represent propositions and propositional functions will be called canonical. With each canonical term, we shall associate a natural number called its degree. This number will represent the number of arguments we expect the propositional function to take. If X is a canonical term of degree k, we write

$$\mathrm{Can}_k(X) .$$

If $k \equiv 0$, we shall interpret X as a proposition.

There are a number of ways in which the canonical terms can be defined, and so we will not settle on one of them now. However, we will assume that they are so defined that the following properties hold:

N1. $\mathrm{Can}_k(X)$ and $X = Y \Rightarrow \mathrm{Can}_k(Y)$;

N2. $\mathrm{Can}_k(X) \Longleftrightarrow \mathrm{Can}_{k+1}(\lambda x.\, X)$;

N3. $\mathrm{Can}_{k+1}(X) \Longleftrightarrow \mathrm{Can}_k(XU)$;

N4. $\mathrm{Can}_1(X)$ and $\mathrm{Can}_1(Y) \Longleftrightarrow \mathrm{Can}_0(\Xi XY)$;

N5. $\mathrm{Can}_1(\mathbf{E})$.

In N3, U is any term. N1, N3 and N4 imply that if the premises of an inference by Rule Ξ or Eq are canonical of degree 0, then so is the conclusion. N5 implies that each axiom (10) is canonical of degree 0. Note, incidentally, that since equality is extensional, N1 and N2 imply the special case of N3 for which U is a variable $\notin X$. Also, from the definitions of Π, $\mathbf{P}$ and $\mathbf{F}$, we get

(14) $\mathrm{Can}_1(X) \Longleftrightarrow \mathrm{Can}_0(\Pi X)$,

(15) $\mathrm{Can}_0(X)$ and $\mathrm{Can}_0(Y) \Longleftrightarrow \mathrm{Can}_0(\mathbf{P}XY)$,

(16) $\mathrm{Can}_1(X)$ and $\mathrm{Can}_1(Y) \Longleftrightarrow \mathrm{Can}_1(\mathbf{F}XY)$.

From here on, we shall use lower case Greek letters for canonical terms, and will assume, for example in '$\vdash \xi X_1 \ldots X_n$', that the degree of ξ is sufficient to make the entire term canonical with degree 0.

Now let us consider the deduction theorem. The result we shall aim at is the following.

Deduction theorem for Ξ. Suppose that $\mathrm{Can}_1(\xi)$, and from the assumption

$$\vdash \xi x$$

and a set B of canonical premises of degree 0 in which x does not occur, there is a proof by rules Ξ and Eq of

$$\vdash \eta .$$

Then there is a proof, from B alone, and in which x does not occur, of

$$\vdash \Xi\xi(\lambda x.\, \eta) .$$

In the notation introduced earlier, this theorem is equivalent to

$$B,\ \xi x \vdash \eta \quad \Rightarrow \quad B \vdash (\xi x \supset_x \eta) ,$$

where x does not occur in the proof on the right. From the theorem we get the usual deduction theorem for **P**, namely

$$(17) \qquad B,\ \zeta \vdash \eta \quad \Rightarrow \quad B \vdash \mathbf{P}\,\zeta\eta ,$$

by putting $\xi \equiv \lambda x.\, \zeta$ for a new variable $x \notin \zeta\eta$, and applying (9).

We can also deduce a 'deduction theorem' for Π, the generalization theorem, which says that if x does not occur in B or ζ, then

$$(18) \qquad B \vdash \zeta x \quad \Rightarrow \quad B \vdash \Pi\zeta .$$

(To deduce this, notice that the hypothesis implies that $B, \mathbf{E}x \vdash \zeta x$, and use (11).)

Now it turns out that the axiom-schemes we need for the deduction-theorem to hold are closely related to the schemes **(FK)** and **(FS)** of Chapter 9. If we write **(FK)** using our current abbreviations, including (12), and a few applications of Rule Eq, we get

$$\vdash \alpha x \supset_x (\beta y \supset_y \alpha(\mathbf{K}xy)), \qquad (x, y \notin \alpha\beta).$$

(Remember that equality is extensional in this chapter, so we have $U = \lambda x. Ux$ for $x \notin U$.) This scheme is equivalent, by Rule Eq, to

$$\vdash \alpha x \supset_x (\beta y \supset_y \alpha x) \qquad (x, y \notin \alpha\beta),$$

or, to use the usual dot notation instead of parentheses,

$$(19) \qquad \vdash \alpha x \supset_x . \beta y \supset_y \alpha x .$$

This is the first axiom-scheme required. The second one, related to **(FS)** but not a direct translation of it, is

$$(20) \qquad \vdash \alpha u \supset_u . \beta uv \supset_v \gamma xuv : \supset_x : \alpha w \supset_w \beta w(yw). \supset_y \alpha z \supset_z \gamma xz(yz) .$$

Here, u, v, w, x, y, z must not occur in $\alpha\beta\gamma$. For comparison, the direct translation of **(FS)** is

$$(21) \qquad \vdash \alpha u \supset_u . \beta v \supset_v \gamma(xuv) : \supset_x : \alpha w \supset_w \beta(yw). \supset_y . \alpha z \supset_z \gamma(xz(yz)) .$$

Finally, besides (19) and (20) we need all the axioms generated by the following rule:

Axiom-generating rule. <u>If $\vdash \eta$ is an axiom and</u> $x \in \eta$, <u>then for any canonical term</u> α <u>of degree one,</u>

$$\vdash \Xi\alpha(\lambda x. \eta)$$

<u>is also an axiom.</u>

This rule applies not only to the substitution-instances of (19) and (20), but also to any other axioms we may have in the system. Thus among the axioms for **E** we will have

$$\vdash \mathbf{E}x \ ;$$

and so the generating rule will give us a new axiom

$$\vdash \Xi\alpha\mathbf{E} \ .$$

The rule then applies to this axiom to give a new one (provided α contains a variable), and so on.

<u>**Lemma 10. 1.**</u> <u>From schemes</u> (19) <u>and</u> (20) <u>follows</u>

$$\vdash \Xi\alpha\alpha \ .$$

Proof. By putting $\mathbf{K}\beta$ for β and $\mathbf{B}(\mathbf{B}\gamma)$ for γ in (20), we get (21). Now (19) and (21) are equivalent to (**FK**) and (**FS**), and from these we can deduce by Rule **F**,

$$\vdash \mathbf{F}\alpha\alpha(\mathbf{SKK}) \ .$$

Then, since $\mathbf{SKK} = \mathbf{I}$, and since $\mathbf{F}\alpha\alpha\mathbf{I} = \Xi\alpha\alpha$ by (12), the result follows by Rule Eq.

<u>**Lemma 10. 2.**</u> <u>It follows from</u> (20) <u>that for</u> u, v $\notin \alpha\beta\gamma$,

$$\alpha u \supset_u (\beta v \supset_v \gamma v), \quad \alpha u \supset_u \beta U \quad \vdash \quad \alpha u \supset_u \gamma U \ .$$

Proof. If we put $\mathbf{K}\beta$ in for β and $\mathbf{K}(\mathbf{K}\gamma)$ in for γ in (20), we get, for u, v, w, x, y, z not occurring in $\alpha\beta\gamma U$,

$$\vdash \alpha u \supset_u . \beta v \supset_v \gamma v : \supset_x : \alpha w \supset_w \beta(yw) . \supset_y . \alpha z \supset_z \gamma(yz) \ .$$

Then by the first premise in the lemma, and Rule Ξ in the form (8), we get

$$\vdash \alpha w \supset_w \beta(yw). \supset_y. \alpha z \supset_z \gamma(yz) ,$$

which is equivalent to

$$(22) \qquad \vdash \alpha u \supset_u \beta(yu). \supset_y. \alpha u \supset_u \gamma(yu) .$$

Now by Rule Eq the second premise is equivalent to

$$\alpha u \supset_u \beta((\lambda u. U)u) ;$$

hence, by Rule Ξ in the form (8), with $\lambda u. U$ substituted for y in (22), we deduce

$$\alpha u \supset_u \gamma((\lambda u. U)u) ,$$

from which the result follows.

Theorem 10. 3. If the only rules are rules Ξ and Eq, and the axioms include (19) and (20) and are closed under the generating rule, then the deduction theorem for Ξ holds.

Proof. By a straightforward induction on the proof of $\vdash \eta$. The generating rule deals with the case that $\vdash \eta$ is an axiom in which x occurs; axiom-scheme (19) in the form

$$\eta \supset_u (\xi x \supset_x \eta)$$

deals with the case that η is an axiom or member of B in which x does not occur; Lemma 10. 1 deals with the case that $\eta \equiv \xi x$. For the two induction cases, we can use Rule Eq and Lemma 10. 2. (See [CLg. II] §15B for details.)

Theorem 10. 3 suggests a natural-deduction formulation of this system. This formulation will have as its rules Rule Eq and the following:

$$\Xi e: \quad \frac{\Xi XY \qquad XU}{YU} \qquad\qquad \Xi i: \quad \frac{\begin{matrix}[\xi] \\ \eta\end{matrix}}{\Xi \xi(\lambda x.\ \eta)}$$

(In Ξi, x must not occur in ξ or in any uncancelled premise.) Its axioms will be all the instances of (10), and any extra axioms that may be in the given formulation, but not instances of (19) and (20). To distinguish between the two formulations, we shall follow Curry and call the given one the A-formulation and the natural deduction one the T-formulation.

Theorem 10.4. $B \vdash^{T} \eta \iff B \vdash^{A} \eta$.

Proof. From left-to-right, the theorem follows by Theorem 10.3. From right-to-left, it is sufficient to prove (19) and (20), and show that if $\vdash X$ is deducible and $x \in X$ but $x \notin \alpha$, then

$$\vdash \Xi\alpha(\lambda x.\ X)$$

is also deducible.

The latter result clearly follows from Rule Ξi.

Scheme (19) can be derived by first proving from the definition of **F** that rules Fe and Fi of Definition 9.16 hold, and then using the proof of Theorem 9.17.

For (20), we have (assuming that x, y, z, u, v, w do not occur in $\alpha\beta\gamma$, and suppressing applications of Rule Eq):

$$\dfrac{\dfrac{\dfrac{\dfrac{\dfrac{\alpha u \supset_u (\beta uv \supset_v \gamma xuv) \qquad \alpha z}{\beta zv \supset_v \gamma xzv}\ \Xi e\ (8) \qquad \dfrac{\alpha w \supset_w \beta w(yw) \qquad \alpha z}{\beta z(yz)}\ \Xi e\ (8)}{\gamma xz(yz)}\ \Xi e\ (8)}{\alpha z \supset_z \gamma xz(yz)}\ \Xi i - 1}{\alpha w \supset_w \beta w(yw) . \supset_y . \ \alpha z \supset_z \gamma xz(yz)}\ \Xi i - 2}{\alpha u \supset_u (\beta uv \supset_v \gamma xuv) :\supset_x : \alpha w \supset_w \beta w(yw) . \supset_y . \ \alpha z \supset_z \gamma xz(yz)}\ \Xi i - 3 \ .$$

This completes the proof.

10C. A GENTZEN L-SYSTEM

In the same way as the natural deduction system for **F** in Section 9B led to an L-formulation, so the T-rules for Ξ suggest the following L-rules. (The same notation conventions are used here as in Section 9C; in particular, 'M' denotes a sequence of canonical terms of degree 0.)

$$*\Xi: \quad \dfrac{\mathbf{M} \Vdash \alpha U \qquad \mathbf{M},\ \beta U \Vdash \gamma}{\mathbf{M},\ \Xi\alpha\beta \Vdash \gamma}$$

$$\Xi*: \quad \dfrac{\mathbf{M},\ \alpha x \Vdash \beta x}{\mathbf{M} \Vdash \Xi\alpha\beta} \qquad (x \text{ not in } \mathbf{M},\ \alpha\beta)\ .$$

Now if we do set up an L-system, the purpose will be to prove the consistency of the original A-formulation, by showing that it is equivalent to the L-formulation and proving the elimination theorem (ET) for the L-formulation. In Chapter 9, ET was proved by induction on the structure of the constituent which was eliminated. But since we have the full equality-rule now in this chapter, we need to look for another way of defining a 'rank' for each constituent, which is invariant of equality. What we therefore do here is to define a special set of terms (called 'canterms' for short), which satisfy the assumptions we have made about canonical terms and which have a rank on which we can make the required induction.

We shall assume that there is a finite or infinite sequence of canonical atoms $\Theta_1, \Theta_2, \ldots,$ each of which is a constant non-redex atom distinct from the connectives and quantifiers. With each canonical atom Θ there is associated a number $dg(\Theta)$, the degree of Θ. The atom **E** is included as a canonical atom with degree 1.

Definition 10. 5. The term ξ is said to be a canonical simplex iff

$$\xi \equiv \Theta U_1 \ldots U_m$$

where Θ is a canonical atom of degree m and $U_1, \ldots, U_m$ are any terms. The term ξ is a proper canterm of rank m and degree n, iff

either (a) ξ is a canonical simplex and $m \equiv n \equiv 0$;

or (b) $\xi \equiv \lambda x.\, \eta$ where η is a proper canterm of rank m and degree $n - 1$;

or (c) $\xi \equiv \Xi\zeta\eta$ where ζ and η are proper canterms of degree 1, and $n \equiv 0$, and $m \equiv 1 + \text{rank}(\zeta) + \text{rank}(\eta)$.

Finally, ξ is a canterm of rank m and degree n iff there is a proper canterm η of rank m and degree n such that $\xi \succ\!\!- \eta$.

Theorem 10. 6. If $\text{Can}_k(X)$ is interpreted to mean that X is a canterm of degree k, then N1-N5 hold.

Proof. Straightforward but tedious; see [CLg. II] §15B3, Theorem 3.

Theorem 10. 7. The degree and rank of every canterm are unique.

Proof. See [CLg. II] §15B3, Theorem 4 and the subsequent discussion.

We are now in a position to define the L-formulation for the given system. The axioms will be all statements of the form

$$\xi \Vdash \xi$$

where ξ is a canterm of degree 0; also all statements

$$\Vdash \mathbf{EX}$$

where **X** is any term, together with axioms corresponding to any extra axioms that the given A-formulation may have. The rules will be the structural rules *C, *K, *W from Chapter 9, where the new constituent introduced by *K is now restricted to be a canterm of degree 0; also two <u>expansion rules</u>

$$\text{*Exp:} \quad \frac{\text{M, Y} \Vdash \text{Z}}{\text{M, X} \Vdash \text{Z}} \qquad\qquad \text{Exp*:} \quad \frac{\text{M} \Vdash \text{Y}}{\text{M} \Vdash \text{X}}$$

where in both rules $\text{X} \succ \text{Y}$; and finally, the two operational rules $*\Xi$ and $\Xi*$ described earlier.

If this definition is adopted, then ET can be proved by induction on the rank of the eliminated constituent. Furthermore, this system can be proved equivalent to the T-formulation just as in Chapter 9. (See [CLg. II] §15B4.)

It is also possible to extend this system to include independent rules for other connectives and quantifiers. This can be done by introducing new nonredex atoms Λ, V, **O** and Σ to represent, respectively, conjunction, disjunction, a standard false proposition, and the existential quantifier. Their corresponding operational rules would be

$$*\Lambda: \quad \frac{\text{M}, \ \xi \Vdash \alpha}{\text{M}, \ \Lambda\xi\eta \Vdash \alpha} \qquad\qquad \frac{\text{M}, \ \eta \Vdash \alpha}{\text{M}, \ \Lambda\xi\eta \Vdash \alpha}$$

$$\Lambda^*: \quad \frac{\mathbf{M} \Vdash \xi \qquad \mathbf{M} \Vdash \eta}{\mathbf{M} \Vdash \Lambda\xi\eta}$$

$$*\mathbf{V}: \quad \frac{\mathbf{M},\ \xi \Vdash \alpha \qquad \mathbf{M},\ \eta \Vdash \alpha}{\mathbf{M},\ \mathbf{V}\xi\eta \Vdash \alpha}$$

$$\mathbf{V}^*: \quad \frac{\mathbf{M} \Vdash \xi}{\mathbf{M} \Vdash \mathbf{V}\xi\eta} \qquad \frac{\mathbf{M} \Vdash \eta}{\mathbf{M} \Vdash \mathbf{V}\xi\eta}$$

$$*\Sigma: \quad \frac{\mathbf{M},\ \eta\mathrm{x} \Vdash \mathbf{Y}}{\mathbf{M},\ \Sigma\eta \Vdash \mathbf{Y}} \qquad \Sigma^*: \quad \frac{\mathbf{M} \Vdash \eta\mathbf{U}}{\mathbf{M} \Vdash \Sigma\eta}\ ,$$

where in $*\Sigma$ the x does not occur in the conclusion. If negation was defined by

$$\neg \equiv \lambda\mathrm{x}.\,\mathbf{PxO}\ ,$$

then it would obey the following rules:

$$\frac{\mathbf{M} \Vdash \eta \qquad \mathbf{M},\ \mathbf{O} \Vdash \alpha}{\mathbf{M},\ \neg\ \eta \Vdash \alpha} \qquad \frac{\mathbf{M},\ \eta \Vdash \mathbf{O}}{\mathbf{M} \Vdash \neg\,\eta}\ .$$

We could then add appropriate clauses to the definition of canterm for these new atoms, and **ET** would still be provable. (See [CLg. II] §15D1.) It is not difficult to give equivalent T- and A-formulations of this extended system.

There are other modifications which can be made to this system without changing these essential results; for details, see [CLg. II] §15D.

One further modification we might like to make is to adjoin an atom **Q** (as a canonical atom of degree 2), and postulate the axiom scheme

$$\Vdash \mathbf{QXX}$$

and the rule

$$*Q: \quad \frac{M \Vdash ZX \qquad M, ZY \Vdash \alpha}{M, QXY \Vdash \alpha} .$$

These correspond to the rules for **Q** discussed earlier, and if ET is provable, the **L**-system here is equivalent to the result of adding the previous rules to the A-formulation of restricted generality. However, it is not clear how ET is to be proved, since the rank of the new constituent **QXY** introduced by rule *Q is always 0, whereas the rank of the corresponding ZX and ZY can be anything.

10D. UNIVERSAL GENERALITY

In this section we shall see how the program of the last two sections can be carried out for a system whose nonredex atoms are Π, **P** and **E** instead of Ξ and **E**, and whose deduction-rules are Rules Π, **P** and Eq, together with the axioms (10) for **E**.

The first thing we must do in this case is to replace N4 in the conditions on canonical terms by (14) and (15). Then the two deduction theorems we want are

Deduction theorem for P. _If_ B, $\xi \vdash \eta$, _then_

B $\vdash$ **P**$\xi\eta$.

Generalization theorem for **Π**. _If_ B $\vdash \eta$x _and_ x _does not occur in_ B _or_ η, _then_

B $\vdash \Pi\eta$

by a derivation in which x _does not occur._

These theorems will both hold if we adjoin to the system the following axiom-schemes.

(23) $\vdash \alpha \supset . \beta \supset \alpha$,

(24) $\vdash \alpha \supset . \beta \supset \gamma {:\supset:} \alpha \supset \beta . \supset . \alpha \supset \gamma$,

(25) $\vdash (\forall x)\alpha x . \supset . \alpha U \qquad (x \notin \alpha)$,

$\vdash \alpha \supset (\forall x)\alpha \qquad (x \notin \alpha)$,

$\vdash (\forall x)(\alpha x \supset \beta x) . \supset . (\forall x)\alpha x \supset (\forall x)\beta x \qquad (x \notin \alpha\beta)$,

and define the set of axioms to be closed under the following rule:

<u>If</u> $\vdash \eta$ <u>is an axiom and</u> $x \in \eta$, <u>then</u> $\vdash \Pi(\lambda x. \eta)$ <u>is an axiom.</u>

Furthermore, we can get a T-formulation similarly to Section B, based on Rule Eq and the following rules:

$$\text{Pe:} \quad \frac{\mathbf{P}\xi\eta \qquad \xi}{\eta} \qquad\qquad \text{Pi:} \quad \frac{\begin{matrix}[\xi] \\ \eta\end{matrix}}{\mathbf{P}\xi\eta}$$

$$\Pi\text{e:} \quad \frac{\Pi\eta}{\eta U} \qquad\qquad \Pi\text{i:} \quad \frac{\eta x}{\Pi\eta} ,$$

where in Rule Πi, x does not occur in η or in any uncancelled premise of the deduction. This T-formulation is equivalent to the T-formulation we had for Ξ, in the sense that if Ξ is defined in terms of Π and **P** as at the start of the chapter, then Rules Ξi and Ξe are deducible from the above rules (but the axioms for **E** are not deducible); and if in the Ξ-based system we define **P** and Π by (9) and (11), then the above rules are derivable.

An alternative system based on Π and **P** can be formulated by adjoining Rule Πi (i. e. the generalization theorem) as a primitive rule to the A-formulation. Then for the deduction theorem for **P** we only need axiom-schemes (23), (24), (25), and

$$\vdash (\forall x)(\alpha \supset \beta x) . \supset . \alpha \supset (\forall x)\beta x \qquad (x \notin \alpha\beta) .$$

Furthermore, this system is equivalent to the above T-formulation. ([CLg. II] §16C1.)

To get an L-formulation for $\mathbf{P}$ and Π we simply replace the rules $*\Xi$ and $\Xi*$ and the axioms for $\mathbf{E}$ in the previous section by the following rules:

$$*\mathbf{P}: \quad \frac{M \Vdash \alpha \qquad M, \beta \Vdash \gamma}{M, \mathbf{P}\alpha\beta \Vdash \gamma} \qquad\qquad \mathbf{P}*: \quad \frac{M, \alpha \Vdash \beta}{M \Vdash \mathbf{P}\alpha\beta}$$

$$*\Pi: \quad \frac{M, \alpha U \Vdash \gamma}{M, \Pi\alpha \Vdash \gamma} \qquad\qquad \Pi*: \quad \frac{M \Vdash \alpha x}{M \Vdash \Pi\alpha} \; ,$$

where in $\Pi*$ the x must not occur in the conclusion. In Definition 10.5 (canterms), we replace (c) by

(c') Either $\xi \equiv \Pi\eta$ where η is a proper canterm of degree 1, and $n \equiv 0$, and $m \equiv 1 + \text{rank}(\eta)$,

or $\xi \equiv \mathbf{P}\zeta\eta$ where ζ and η are proper canterms of degree 0, and $n \equiv 0$, and $m \equiv 1 + \text{rank}(\zeta) + \text{rank}(\eta)$.

Then ET is provable, and from it the equivalence of the L and T-formulations. And just as in the previous section, we can adjoin the other connectives and quantifiers with their corresponding rules, without affecting ET.

Now the systems we have been dealing with so far impose restrictions on the deduction theorem for Π as well as $\mathbf{P}$. On the other hand, Curry's paradox only indicated that the restriction was needed for P. If Ξ is primitive, the restriction on Π cannot be avoided without also eliminating the restrictions for $\mathbf{P}$. But if Π and $\mathbf{P}$ are the atoms, then we can arrange the postulates so that the restriction is imposed only on the deduction theorem for P. We can do this, for example, by taking a T-formulation in which Rule Πi is re-stated as follows:

$$\frac{Xx}{\Pi X}$$

where it is not assumed that X is canonical. The rules for P will be unaltered. An L-formulation of this system can be constructed and used to prove that if Rule Pe is omitted or restricted, then the system is consistent, even when rules for Λ, V, Σ are adjoined without canonicalness restrictions.

This is done as follows. Let the axioms be

$$X \Vdash X$$

for all terms X. Let the rules be the structural rules *C, *W, and *K without any restriction on the new constituent introduced; also the expansion rules, the operational rules for Π, Σ, Λ, V, and the Rule *P (but not P*), where there are now no restrictions about the canonicalness of the constituents of these rules. We may also postulate the Rule *Q and the axiom-scheme $\Vdash$ **QXX**.

It is not at all difficult to show that this system is equivalent to the T-formulation considered above, provided that ET holds. And perhaps most surprisingly, it is possible to show that ET does hold, and hence that the T-formulation is consistent. For the proof of ET, see [CLg. II] §16C2. This proof still goes through if we adjoin the axiom-scheme

$$\Vdash \mathbf{PXX} \qquad \text{(for all X)},$$

and the restricted rule P*:

$$\frac{M, X \Vdash Y}{M \Vdash \mathbf{PXY}}$$

in which no 'ancester' ([CLg. II] §12C3) of X in the proof of

$M, X \Vdash Y$ is introduced by an operational rule on the left (i. e. $*\Lambda$, $*V$, $*\Sigma$, $*P$, $*\Pi$). See [CLg. II] §16C3. One condition which guarantees this restriction is that the head of **X** be a nonredex atom distinct from all the connectives and quantifiers.

10E. OTHER DEVELOPMENTS

Remark 10. 8. The A-formulations we considered in Sections B and D involved an infinity of axioms, since each of (19), (20), (23), etc. was an axiom-scheme. It is possible, however, to get a formulation based on only a finite number of axioms (excepting the axioms for **E**).

To do this, we first adjoin a new atom **H**, and define

$$\mathbf{H}_0 \equiv \mathbf{H}\,,$$
$$\mathbf{H}_{n+1} \equiv \mathbf{FEH}_n\,.$$

Then if $\mathrm{Can}_k(\mathbf{X})$ is interpreted as

$$(26) \qquad \vdash \mathbf{H}_k\mathbf{X}\,,$$

the canonicalness properties N1-N3 will hold. The other properties, N4 and N5, can be given to (26) by suitable axioms. With this interpretation of canonicalness, we can now formally quantify over canonical terms and replace the previous axiom-schemes by axioms; for example, the scheme (19) can be replaced by the axiom

$$\vdash H_1 u \supset_u . H_1 v \supset_v (ux \supset_x . vy \supset_y ux)\,.$$

The axioms for **E** can also be simplified: instead of an axiom $\vdash \mathbf{EX}$ for each **X**, we can postulate $\vdash \mathbf{EX}$ only when **X** is an atom, and deduce it for composite terms by the axiom

(27) $\vdash \mathbf{E}x \supset_x . \mathbf{E}y \supset_y \mathbf{E}(xy)$.

(In a λ-system, (27) alone is not enough; but if we adjoin three extra axioms $\vdash \mathbf{E}(\mathbf{S}_\lambda)$, $\vdash \mathbf{E}(\mathbf{K}_\lambda)$ and $\vdash \mathbf{E}(\mathbf{I}_\lambda)$, then for each $\mathbf{X}$ we shall have $\mathbf{X} = \mathbf{X}_{H\lambda}$ from Theorem 6. 3(d), and $\vdash \mathbf{EX}_{H\lambda}$ will hold by (27) and the three axioms, so $\vdash \mathbf{EX}$ will hold by Rule Eq.)

The only deduction-rules needed in this system would be Rules Ξ (or Π and **P**) and Eq, and the following (either as a derived or postulated rule):

$$\mathbf{X} \vdash \mathbf{HX} .$$

For further information, see [CLg. II] §15C.

In [CLg. II] §17C3 it is pointed out that systems like this are related to type-theories in which **E** is included as a basic type.

Remark 10. 9. An interesting way of extending type-theory is to use a basis, instead of **F**, a notion of type in which the type of the ith argument of an n-argument function depends on the values of the previous i - 1 arguments. In terms of Ξ, the statement

(28) $\vdash \xi_1 x_1 \supset_{x_1} . \xi_2 x_1 x_2 \supset_{x_2} \eta x_1 x_2 (Yx_1 x_2)$

asserts that Y represents a two-place function with this property, for given ξ_1, ξ_2, η. Because if $\mathbf{X}_1$ represents an object of type ξ_1 (i. e. $\vdash \xi_1 \mathbf{X}_1$ holds), then by rule Ξ in the form (8),

$$\vdash (\xi_2 \mathbf{X}_1) x_2 \supset_{x_2} (\eta \mathbf{X}_1 x_2)(Y\mathbf{X}_1 x_2) ,$$

which means that the type of the arguments $\mathbf{X}_2$ that are acceptable for Y depends on $\mathbf{X}_1$.

The generalization of (28) to n arguments is

$$\vdash \xi_1 x_1 \supset_{x_1} . \xi_2 x_1 x_2 \supset_{x_2} . \ \ldots \ . \xi_n x_1 \ldots x_n \supset_{x_n} \eta x_1 \ldots x_n (Yx_1 \ldots x_n) .$$

If we can define a sequence of terms $\mathbf{G}_n$ so that this statement is equal to

$$\vdash \mathbf{G}_n \, \xi_1 \ldots \xi_n \eta Y ,$$

then we shall have a means of representing these extended types corresponding to the '$\mathbf{F}_n$' notation in Chapters 8 and 9. A suitable definition of $\mathbf{G}_n$ is given by

$$\begin{aligned} \mathbf{G}_1 &\equiv \lambda uvy.\, \Xi u(\mathbf{S}vy) \\ &= \lambda uvy.\, ux \supset_x vx(yx) , \\ \mathbf{G}_{n+1} &\equiv \lambda u_1 \ldots u_{n+1} y.\, \mathbf{G}_1 u_1 (\lambda x.\, \mathbf{G}_n (u_1 x) \ldots (u_n x)(yx)) . \end{aligned}$$

(Contrast $\mathbf{G}_1$ with (12) near the end of section A, in which $\mathbf{F}$ is defined in terms of Ξ.) Then from rule Ξ we can deduce the following rule for $\mathbf{G}_1$:

$$\mathbf{G}_1 UVY, \; UX \vdash VX(YX) .$$

An alternative approach is to take $\mathbf{G}_1$ as primitive, the above rule as a postulate, and to develop a theory similar to that developed for $\mathbf{F}$ in Chapter 9. The resulting theory could be used to build up a generalized logical type-theory in which the type of the ith argument of a function may depend on the previous arguments. Such a theory has not yet been attempted, but a few preliminary results are contained in [CLg. II] §15A8.

Remark 10.10. Finally, an interesting subject which is still being developed is the formulation of set theory in terms of combinatory logic; for details, see Bunder [STB].

11. Gödel's Functions of Finite Type

In the past few years there has been some interest in 'proving' the consistency of certain intuitionist formal theories by interpreting their formulae as $\exists\forall$-statements about a class of functions (which depends on the given theory). The original idea came from Gödel [BNN], where he proved the consistency of first-order intuitionist arithmetic using the class of functions to be described in this chapter. By means of a larger class, Spector [PRA] extended the idea to a formalization of intuitionist analysis.

One very natural way of defining these functions is through corresponding classes of λ- or combinatory terms (see Tait [ILT], [IIF], Grzegorczyk [ROA], and Sanchis [FDR]). In this chapter the functions used by Gödel will be defined using combinators with types, and his consistency-proof for arithmetic will be outlined.

Types are defined here as in Definition 8.1, but with only one basic type, $\mathbf{N}$ (for the set of all natural numbers).

Gödel's functions of finite type (often called 'primitive recursive functionals of finite type') may be defined, roughly speaking, as the smallest set of functions which contains the successor-function σ, the number 0 (as a 'function' of no arguments), and is closed under the operations of λ-abstraction, application of a function to an argument, and primitive recursion. We shall not need to make this more precise, as the consistency-proof does not formally involve the functions themselves, but only a system of notations for them. This can be provided as follows.

Definition 11.1. Typed terms are defined as in Definition 8.2, but based on the following atoms only:

(i) $\mathbf{S}_{\alpha\beta\gamma}$, $\mathbf{K}_{\alpha\beta}$, $\mathbf{I}_{\alpha}$ and variables as in Definition 8.2;

(ii) two constants $\bar{0}$, $\bar{\sigma}$ with types $\mathbf{N}$, $\mathbf{FNN}$ respectively;

(iii) for each type α a distinct constant $\mathbf{R}_{\alpha}$ with type

$$\mathbf{F}_3\alpha(\mathbf{F}_2\mathbf{N}\alpha\alpha)\mathbf{N}\alpha .$$

Notation. We shall use the conventions stated after Definition 8.2. Sometimes, when no confusion will result, an $\mathbf{S}_{\alpha\beta\gamma}$ will simply be called '**S**'; similarly for $\mathbf{K}_{\alpha\beta}$, $\mathbf{I}_{\alpha}$ and $\mathbf{R}_{\alpha}$. Terms of form $\bar{\sigma}^n\bar{0}$ will be called numerals, and we define

$$\bar{n} \equiv \bar{\sigma}^n\bar{0} .$$

(Each numeral has type **N**.) Abstraction is defined as in Definition 8.6.

Definition 11.2. Reduction is defined by adding to Definition 8.4 the extra axiom-schemes

$$(11.3)\qquad \begin{aligned} &\mathbf{R}_{\alpha}X^{\alpha}Y^{\mathbf{FN}(\mathbf{F}\alpha\alpha)}\bar{0} \triangleright X, \\ &\mathbf{R}_{\alpha}X^{\alpha}Y^{\mathbf{FN}(\mathbf{F}\alpha\alpha)}(\bar{\sigma}\bar{n}) \triangleright Y\bar{n}(\mathbf{R}_{\alpha}XY\bar{n}) . \end{aligned}$$

(Cf. (3.11).) Equality is defined by the same clauses as reduction, together with the rule

$$X = Y \;\Rightarrow\; Y = X .$$

An R-redex is any term of form $\mathbf{R}XY\bar{n}$. A term is said to be in normal form iff it contains no redexes of any kind.

Theorem 11.4. *If $X = Y$, then X and Y have the same type.*

Proof. By induction on Definition 11.2. The type given to $\mathbf{R}_\alpha$ ensures that both sides of (11.3) have type α. The other axioms and rules are dealt with as in Theorem 8.5.

The above reduction can easily be shown to have all the usual properties (2.4, 2.8, 2.9, 2.12, 2.16, 2.18, 2.20, 2.21 and their corollaries). The following additional property is needed for the consistency-proof.

Theorem 11.5. Every typed term has a unique normal form.

Proof. The uniqueness comes from Rosser's theorem (Corollary 2.18.1 in fact). For the proof of existence, see Sanchis [FDR] §4 or Tait [IIF] §1-2.

All the consistency-proof except this theorem can be carried out in first-order arithmetic; hence by Gödel's well-known result, the proof of this normal-form theorem cannot be formalized as a proof in arithmetic (unless arithmetic is actually inconsistent). Sanchis' proof uses an induction with infinitely many premises. In Schütte [TFE] there is a proof for an equivalent system using induction up to the ordinal number ε_0, based on a proof by W. Howard for λ-conversion. (ε_0 is the least ordinal number ε for which $\omega^\varepsilon = \varepsilon$; Gentzen showed that for each $\alpha < \varepsilon_0$, a proof of induction up to α could be formalized in arithmetic, so induction up to ε_0 is a 'minimal' non-arithmetical principle.) By contrast, the corresponding normal-form theorems in Chapters 8 and 9 have proofs which are finitary.

Corollary 11.5.1. Equality of typed terms is decidable.

Corollary 11.5.2. Every constant term with type **N** reduces to a numeral.

Proof. By the theorem, we need only prove that every irreducible constant term **X** with type **N** is a numeral. This is done by induction on X. First, every X has the form

$$X \equiv bX_1 \dots X_n \qquad (n \geq 0)$$

where b is an atom. It is an easy exercise to show that b cannot be **S**, **K**, **I** or **R**, and hence must be $\overline{0}$ or $\overline{\sigma}$. The result then follows by the induction-hypothesis.

Definition 11. 6. Let ϕ be an n-argument total function of natural numbers. Then a typed term X <u>combinatorially defines</u> ϕ iff

(i) X has type $\mathbf{F}_n \alpha_1 \dots \alpha_n \mathbf{N}$ $\qquad (\alpha_i \equiv \mathbf{N}$ for $i = 1, \dots, n)$.

(ii) $X\overline{m}_1 \dots \overline{m}_n \rhd \overline{\phi(m_1, \dots, m_n)}$ for all $m_1, \dots, m_n$.

(Partial functions are not considered in this definition because $X\overline{m}_1 \dots \overline{m}_n$ reduces to a numeral for all $m_1, \dots, m_n$, and hence such functions could only be definable in the weak sense (cf. Definition 3. 2).

Theorem 11. 7. <u>Every primitive recursive function ϕ can be combinatorially defined by a typed term $\overline{\phi}$.</u>

Proof. In the construction of $\overline{\phi}$ in Theorem 3. 4, replace '**SB**' by '$\overline{\sigma}$' and '$\mathbf{Z}_0$' by '$\overline{0}$' in parts I and II, and replace the constructed **R** by the typed atom $\mathbf{R}_{\mathbf{N}}$ (cf. Remark 3. 10). Give all the variables the type **N**. Then in all the cases the resulting $\overline{\phi}$ has type

$$\mathbf{F}_n \alpha_1 \dots \alpha_n \mathbf{N} \qquad (\alpha_i \equiv \mathbf{N} \text{ for } i = 1, \dots, n),$$

and the proof of Theorem 3. 4 gives

$$\overline{\phi}\,\overline{m}_1 \dots \overline{m}_n \rhd \overline{\phi(m_1, \dots, m_n)}.$$

Remark. This theorem does not extend to all recursive total functions, because if all such functions were definable, the proofs of Theorem 4.5 and Corollary 4.5.3 would imply that equality is undecidable, contrary to Corollary 11.5.1.

Corollary 11.7.1. There exist terms $\overline{+}$, $\overline{\times}$, $\overline{\dot{-}}$ with type $\mathbf{F}_2\mathbf{NNN}$, such that

$$\overline{+}\,\overline{m}\,\overline{n} \rhd \overline{(m+n)},$$

$$\overline{\times}\,\overline{m}\,\overline{n} \rhd \overline{(m \times n)}$$

$$\overline{\dot{-}}\,\overline{m}\,\overline{n} \rhd \overline{(m \dot{-} n)},$$

where $m \dot{-} n \equiv m - n$ if $m \geq n$, but 0 if $m < n$. There is also a term E with type $\mathbf{F}_2\mathbf{NNN}$ such that

$$E\,\overline{m}\,\overline{n} \rhd \overline{0} \text{ if } m \equiv n,$$
$$\overline{1} \text{ if } m \not\equiv n.$$

Proof. Compare Exercise 1, Chapter 3.

Definition 11.8. Corresponding to the propositional connectives $\&$, $\mathbf{V}$, $\supset$, $\neg$, we define terms $\overline{\&}$, $\overline{\mathbf{V}}$, $\overline{\supset}$, $\overline{\neg}$ as follows:

$$\overline{\&} \equiv \overline{+};$$

$$\overline{\neg} \equiv [x^{\mathbf{N}}](\overline{\dot{-}}\,\overline{1}\,x);$$

$$\overline{\supset} \equiv [x^{\mathbf{N}}, y^{\mathbf{N}}]\,\overline{\neg}\,(\overline{\&}\,x(\overline{\neg}y));$$

$$\overline{\mathbf{V}} \equiv [x^{\mathbf{N}}, y^{\mathbf{N}}]\overline{\neg}(\overline{\&}\,(\overline{\neg}x)(\overline{\neg}y)).$$

From this definition, if X and Y are constant terms with type $\mathbf{N}$, then $\overline{\&}XY$, $\overline{\neg}X$, $\overline{\supset}XY$ and $\overline{\mathbf{V}}XY$ will also be constant terms with type $\mathbf{N}$; hence each one must reduce to $\overline{0}$ or to a numeral of form $\overline{k+1}$. And we have

$\bar{\&}XY \rhd \bar{0}$ iff $X \rhd \bar{0}$ and $Y \rhd \bar{0}$;

$\bar{\neg}X \rhd \bar{0}$ iff $X \rhd \overline{k+1}$ for some k;

$\bar{V}XY \rhd \bar{0}$ iff $X \rhd \bar{0}$ or $Y \rhd \bar{0}$;

$\bar{\supset}XY \rhd \bar{0}$ iff $Y \rhd \bar{0}$ or $X \rhd \overline{k+1}$ for some k.

Notation. $\bar{\&}XY$ will be denoted by '$X \mathbin{\bar{\&}} Y$' from now on, and similarly for $\bar{V}$, $\bar{\supset}$, $\bar{\neg}$, $\bar{+}$, $\bar{\times}$, $\bar{\dot{-}}$.

Lemma 11.9. For each α there is a term D_α such that

$$D_\alpha X^\alpha Y^\alpha \bar{0} = X^\alpha$$

$$D_\alpha X^\alpha Y^\alpha \overline{(k+1)} = Y^\alpha .$$

Proof. Compare D^* in Remark 3.10.

This completes the basic combinatory apparatus for the consistency-proof. We now begin the proof proper.

Definition 11.10. Intuitionist arithmetic. The first-order theory of intuitionist arithmetic may be defined as follows (cf. Kleene [IMM] §17-19 and p. 101, whose notational conventions will be used here, except that script letters will be used for formulae).

Arithmetical terms: An infinity of variables are terms, together with one constant, 0; if s and t are terms, then so are (s) + (t), (s). (t), (s)'.

Arithmetical formulae: For any terms s and t, (s) = (t) is a formula; if $\mathscr{A}$ and $\mathscr{B}$ are formulae, then so are $(\mathscr{A})\supset(\mathscr{B})$, $(\mathscr{A})\&(\mathscr{B})$, $(\mathscr{A})\vee(\mathscr{B})$, $\neg(\mathscr{A})$, $\forall x(\mathscr{A})$ and $\exists x(\mathscr{A})$, where x is any variable.

Axiom-schemes:

$(\mathscr{A} \vee \mathscr{A}) \supset \mathscr{A}$,	$\neg(x' = 0)$,
$\mathscr{A} \supset (\mathscr{A} \vee \mathscr{A})$,	$(x = y) \supset (x' = y')$,
$(\mathscr{A} \vee \mathscr{B}) \supset (\mathscr{B} \vee \mathscr{A})$,	$(x' = y') \supset (x = y)$,

$(\mathscr{A} \mathbin{\&} \mathscr{B}) \supset (\mathscr{B} \mathbin{\&} \mathscr{A})$,

$\mathscr{B} \supset (\mathscr{A} \vee \mathscr{B})$,

$(\mathscr{A} \mathbin{\&} \mathscr{B}) \supset \mathscr{B}$,

$(0 = 0') \supset \mathscr{A}$,

$\neg \mathscr{A} \supset (\mathscr{A} \supset (0 = 0'))$,

$(\mathscr{A} \supset (0 = 0')) \supset \neg \mathscr{A}$.

$(x=y)\supset((x=z)\supset(y=z))$,

$x + 0 = x$,

$x + y' = (x + y)'$,

$x \,.\, 0 = 0$,

$x \,.\, y' = (x \,.\, y) + x$.

Deduction-rules:

$$\frac{\mathscr{A},\ \mathscr{A} \supset \mathscr{B}}{\mathscr{B}} \qquad \frac{\mathscr{A} \supset \mathscr{B},\ \mathscr{B} \supset \mathscr{C}}{\mathscr{A} \supset \mathscr{C}}$$

$$\frac{\mathscr{A} \supset (\mathscr{B} \supset \mathscr{C})}{(\mathscr{A} \mathbin{\&} \mathscr{B}) \supset \mathscr{C}} \qquad \frac{(\mathscr{A} \mathbin{\&} \mathscr{B}) \supset \mathscr{C}}{\mathscr{A} \supset (\mathscr{B} \supset \mathscr{C})}$$

$$\frac{\mathscr{B} \supset \mathscr{C}}{(\mathscr{A} \vee \mathscr{B}) \supset (\mathscr{A} \vee \mathscr{C})} \qquad \frac{\forall x\, \mathscr{A}(x)}{\mathscr{A}(t)}$$

$$\frac{\mathscr{C} \supset \mathscr{A}(x)}{\mathscr{C} \supset \forall x\, \mathscr{A}(x)} \qquad \frac{\mathscr{C} \supset \forall x\, \mathscr{A}(x)}{\mathscr{C} \supset \mathscr{A}(x)}$$

$$\frac{\mathscr{A}(x) \supset \mathscr{C}}{\exists x\, \mathscr{A}(x) \supset \mathscr{C}} \qquad \frac{\exists x\, \mathscr{A}(x) \supset \mathscr{C}}{\mathscr{A}(x) \supset \mathscr{C}}$$

$$\frac{\mathscr{A}(0), \qquad \forall x(\mathscr{A}(x) \supset \mathscr{A}(x'))}{\mathscr{A}(x)} \ .$$

(In the lower four quantifier-rules, $\mathscr{C}$ must not contain x free, and in the first quantifier-rule, t is any term free for x in $\mathscr{A}(x)$.)

These axioms and rules are designed especially to make the consistency-proof easy. Their equivalence, for example, to the rules in Kleene [IMM] pp. 82 and 101 is left as a tedious exercise. (Compare Spector [PRA] pp. 3-4.)

We shall interpret each arithmetical formula as an $\exists\forall$-formula of the following system, which is just a formalization of the theory of reduction, with quantifiers added.

Definition 11.11 (A formal theory of reduction).

Terms: All the typed combinatory terms of Definition 11.1.

Formulae: If X and Y are terms, then the result, $X \rhd Y$, of placing a symbol $\rhd$ between X and Y, is a formula. If $\mathscr{A}$ is a formula and $x_1, \ldots, x_n$ are variables of any type, then

$$\forall x_1, \ldots, x_n(\mathscr{A}), \qquad \exists x_1, \ldots, x_n(\mathscr{A})$$

are formulae.

Axiom-schemes and deduction-rules: see Definition 11.2.

There are no rules for the quantifiers.

We shall be principally interested in formulae of form $\exists x_1, \ldots, x_m \forall y_1, \ldots, y_n(X \rhd \bar{0})$; in such formulae the term X may be called the matrix.

Definition 11.12. Let X have type **N**. Then a formula of form $X \rhd \bar{0}$, whose variables are $v_1, \ldots, v_n$, is said to be true iff for all constant terms $V_1, \ldots, V_n$ with the same types as $v_1, \ldots, v_n$ respectively,

$$[V_1/v_1]\ldots[V_n/v_n]X \rhd \bar{0}$$

is provable in the theory of reduction.

A formula of form $\exists y_1, \ldots, y_m \forall z_1, \ldots, z_n(X \rhd \bar{0})$, whose free variables are $x_1, \ldots, x_k$, is said to be true iff there exist constant terms $Y_1, \ldots, Y_m$ such that the formula

$$[(Y_1 x_1 \ldots x_k)/y_1]\ldots[(Y_m x_1 \ldots x_k)/y_m]X \rhd \bar{0}$$

is true in the previous sense (and $Y_i x_1 \ldots x_k$ has the same type as y_i, for each $i = 1, \ldots, m$).

Definition 11.13. To each arithmetical term t we assign a corresponding combinatory term t^*, by induction on t as follows.

(i) To each variable v, assign a distinct type-**N** variable v^*;

(ii) $0^* \equiv \overline{0}$,

(iii) $(t_1 + t_2)^* \equiv t_1^* \mathbin{\overline{+}} t_2^*$,

(iv) $(t_1 \cdot t_2)^* \equiv t_1^* \mathbin{\overline{\times}} t_3^*$,

(v) $(t')^* \equiv \overline{\sigma} t^*$.

Definition 11.14. To each arithmetical formula $\mathscr{F}$, whose free variables are $x_1, \ldots, x_k$, we assign a formula $\mathscr{F}^*$ from Definition 11.11, with the form

$$\mathscr{F}^* \equiv \exists y_1, \ldots, y_m \forall z_1, \ldots, z_n (\mathbf{X} \rhd \overline{0}) ;$$

$\mathscr{F}^*$ is defined by induction on $\mathscr{F}$, as follows.

(i) If $\mathscr{F}$ is $(t_1) = (t_2)$, define

$$(t_1 = t_2)^* \equiv ((\mathrm{E}\, t_1^* \, t_2^*) \rhd \overline{0}) .$$

(ii) Suppose $\mathscr{F}$ is $\mathscr{A} \,\&\, \mathscr{B}$, $\mathscr{A} \supset \mathscr{B}$, $\mathscr{A} \vee \mathscr{B}$, $\neg \mathscr{A}$, $\exists x \mathscr{A}$ or $\forall x \mathscr{A}$, and we have already defined

$$\mathscr{A}^* \equiv \exists y_1, \ldots, y_m \forall z_1, \ldots, z_n (\mathrm{A} \rhd \overline{0}) ,$$
$$\mathscr{B}^* \equiv \exists v_1, \ldots, v_p \forall w_1, \ldots, w_p (\mathrm{B} \rhd \overline{0}) .$$

Then we first make sure that all the quantified variables in $\mathscr{A}^*$ and $\mathscr{B}^*$ are distinct, by replacing them by new ones if necessary. In what follows, for ease of reading we shall only treat the case $m \equiv n \equiv q \equiv p \equiv 1$. Also '$\rhd \overline{0}$' will be omitted from $\mathscr{A}^*$ and $\mathscr{B}^*$, and the relevant variables displayed after A and B, so that the above identities read as

$$\mathscr{A}^* \equiv \exists y \forall z \mathrm{A}\{y, z\}, \qquad \mathscr{B}^* \equiv \exists v \forall w \mathrm{B}\{v, w\}.$$

Then $\mathscr{F}^*$ is defined as follows.

(a) $(\mathscr{A} \,\&\, \mathscr{B})^* \equiv \exists y, v \forall z, w \quad (A\{y, z\} \mathbin{\bar{\&}} B\{v, w\})$.

(b) $(\mathscr{A} \vee \mathscr{B})^* \equiv \exists y, v, d \forall z, w \quad (((Ed\bar{0}) \mathbin{\bar{\&}} A\{y, z\}) \mathbin{\bar{\vee}}$ $((Ed\bar{1}) \mathbin{\bar{\&}} B\{v, w\}))$, where d is a new type-**N** variable.

(c) $(\neg \mathscr{A})^* \equiv \exists z' \forall y \mathbin{\bar{\neg}} A\{y, z'y\}$ where z' is a new variable with type $\mathbf{F}\alpha\beta$, where α, β are the types of y, z.

(d) $(\mathscr{A} \supset \mathscr{B})^* \equiv \exists z', v' \forall y, w \quad (A\{y, z'yw\} \mathbin{\bar{\supset}} B\{v'y, w\})$, where z', v' are new variables whose types are such that $z'yw$, $v'y$ have the same types as z, v respectively.

(e) $(\exists x \mathscr{A})^* \equiv \exists x^*, y \forall z \quad A\{y, z\}$.

(f) $(\forall x \mathscr{A})^* \equiv \exists y' \forall x^*, z \quad A\{y'x^*, z\}$, where y' has type $\mathbf{FN}\beta$, where β is the type of y.

Remark 11.15 (Informal motivation). Suppose we use the logical symbols and variables informally for the moment, and assume that $\forall x \exists y \mathscr{A}(x, y)$ is only true if there is a function y' which assigns to each x a value of y satisfying A; that is,

$$(11.16) \qquad \forall x \exists y \mathscr{A}(x, y) \iff \exists y' \forall x \mathscr{A}(x, y'(x)) .$$

Then $\mathscr{F}^*$ turns out to be equivalent to $\mathscr{F}$.

Case (i): $\mathscr{F}$ is $(t_1) = (t_2)$. $\mathscr{F}$ is true for all substitutions of numbers for its variables if and only if $E t_1^* t_2^*$ reduces to $\bar{0}$ for all substitutions of type-**N** constants for its variables.

Case (ii): suppose that, informally, we are given

$$\mathscr{A} \iff \exists y \forall z (A\{y, z\} \rhd \bar{0}) ,$$
$$\mathscr{B} \iff \exists v \forall w (B\{v, w\} \rhd \bar{0}) .$$

Then we can deduce $\mathscr{F} \iff \mathscr{F}^*$ as follows.

(a): $\exists y\ \forall z(A \triangleright \bar{0})\ \&\ \exists v\ \forall w(B \triangleright \bar{0})$

$\Longleftrightarrow\ \exists y, v\quad \forall z, w\quad (A \triangleright \bar{0}\ \&\ B \triangleright \bar{0})$

$\Longleftrightarrow\ \exists y, v\quad \forall z, w\quad ((A\ \bar{\&}\ B) \triangleright \bar{0})$.

(b): $\exists y\ \forall z(A \triangleright \bar{0}) \vee \exists v\ \forall w(B \triangleright \bar{0})$

$\Longleftrightarrow\ \exists d_1(d_1 \equiv 0\ \&\ \exists y\ \forall z(A \triangleright \bar{0}).\vee.\ d_1 \not\equiv 0\ \&\ \exists v\ \forall w(B \triangleright \bar{0}))$

$\Longleftrightarrow\ \exists y, v, d_1\quad \forall z, w\quad (d_1 \equiv 0\ \&\ A \triangleright \bar{0}.\vee.\ d_1 \not\equiv 0\ \&\ B \triangleright \bar{0})$

$\Longleftrightarrow\ \exists y, v, d\quad \forall z, w\quad (d \equiv 0\ \&\ A \triangleright \bar{0}. \vee d \equiv 1\ \&\ B \triangleright \bar{0})$

since if d_1 exists, a suitable d is $1 \dot{-} (1 \dot{-} d_1)$. The rôle of d can be seen by working through the proof of Theorem 11.17 later.

(c): $\neg\ \exists y\ \forall z(A\{y, z\} \triangleright \bar{0})$

$\Longleftrightarrow\ \forall y\ \exists z(\bar{\neg}\ A\{y, z\} \triangleright \bar{0})$

$\Longleftrightarrow\ \exists z'\ \forall y(\bar{\neg}\ A\{y, z'y\} \triangleright \bar{0})$ by (11.16).

(d): $\exists y\ \forall z(A\{y, z\} \triangleright \bar{0}) \supset \exists v\ \forall w(B\{v, w\} \triangleright \bar{0})$

$\Longleftrightarrow\ \forall y[\ \forall z(A \triangleright \bar{0}).\supset.\ \exists v\ \forall w(B \triangleright \bar{0})]$

$\Longleftrightarrow\ \forall y\ \exists v[\ \forall z(A \triangleright \bar{0}).\supset.\ \forall w(B \triangleright \bar{0})]$

$\Longleftrightarrow\ \forall y\ \exists v\ \forall w\ \exists z[A\{y, z\} \triangleright \bar{0}.\supset.B\{v, w\} \triangleright \bar{0}]$

$\Longleftrightarrow\ \exists v', z'\quad \forall y, w\quad [A\{y, z'yw\} \triangleright \bar{0}.\supset.B\{v'y, w\} \triangleright \bar{0}]$,

using (11.16) three times for the last equivalence.

(e): trivial.

(f): from (11.16).

Theorem 11.17. If $\mathcal{F}$ is provable in intuitionist first-order arithmetic, then $\mathcal{F}^*$ is true in the sense of Definition 11.12.

Proof. By induction on the proof of $\mathcal{F}$; see Appendix 2.

Theorem 11.18. Intuitionist first-order arithmetic is consistent.

Proof. Let $\mathcal{F}$ be the equation $0' = 0$; then $\mathcal{F}^*$ is

$E\bar{1}\bar{0} \triangleright \bar{0}$

which is not true formula, because $E\bar{1}\bar{0}$ reduces to $\bar{1}$, and a term cannot have two normal forms. Hençe by Theorem 11.17, this $\mathscr{F}$ is not provable.

Remarks. More information could be got out of this consistency- proof by substituting 'provable' (in a suitably-defined extension of the theory of reduction in Definition 11.11) for 'true' in Theorem 11.17; compare Spector [**PRA**].

Finally, although it seemed most natural to base this chapter on the Chapter 8 approach to types, the Chapter 9 approach could easily have been used instead.

Appendix 1
Proof of the Church-Rosser Theorem

The Church-Rosser theorem (Theorem 1.8) states that if $U \triangleright X$ and $U \triangleright Y$, then there is a Z such that $X \triangleright Z$ and $Y \triangleright Z$. The original proof by Church and Rosser is given in Church [CLC] p. 25 Theorem 7 **XXVII**, and is analysed in detail in Curry and Feys [CLg. I] Chapter 4. The proof below is adapted from a proof by P. Martin-Löf and W. W. Tait, which is much simpler than previous proofs. *

Recall that reduction is defined by series of replacements with form

(i) $(\lambda x. M)N$ may be replaced by $[N/x]M$,

(ii) $(\lambda y. Z)$ may be replaced by $\lambda v. [v/y]Z$ if y is not bound in Z and v is neither free nor bound in Z.

We shall call terms $(\lambda x. M)N$ β-redexes, and terms $\lambda y. Z$ restricted as in (ii) α-redexes. An α-contraction will be a replacement (ii), and a β-contraction will be a replacement (i). If a series of α-contractions changes X to Y, we shall say that X α-reduces to Y ($X \triangleright_\alpha Y$). Similarly for X β-reduces to Y ($X \triangleright_\beta Y$).

To simplify the notation, the work 'redex' will often be used to mean a particular occurrence of a redex in a term.

Definition 1 (Residuals). Let U be a term containing occurrences P, Q of β-redexes, such that Q does not contain P. Let U' be the result of contracting Q in U. We define the residual of P in U' to be an occurrence of a redex in U', determined as follows.

Case 1: P, Q are non-overlapping parts of U. Then contracting Q leaves P unchanged in U'. This occurrence of P is the residual of P.

* For Martin-Löf's form of the proof, see his 'A Theory of Types', to be published, or Barendregt [SEM] p. 128.

Case 2: P is the same part of U as Q is. Then contracting Q is the same as contracting P. There is no residual of P.

Case 3: Q is part of P. Then P has form $(\lambda v. Y)Z$ and Q is in Y or in Z. Then contracting Q changes P to a term of form $(\lambda v. Y')Z$ or $(\lambda v. Y)Z'$. This is taken as the residual of P.

Definition 2. Given a term X, we define special reductions of X as follows. Let $R_1, \ldots, R_n$ be any set of occurrences of β-redexes in X. Contract any R_i which contains no other R_j. (Suppose $i = 1$ for example.) This leaves $n - 1$ residuals $R'_2, \ldots, R'_n$, of $R_2, \ldots, R_n$. Contract any R'_j which contains no other R'_k. This leaves $n - 2$ residuals. Repeat this process as often as desired, or until there are no residuals left. Then after the β-steps, perform some α-steps (perhaps none). If Y is the result of a reduction of this form, we shall say

$$X \rhd_s Y .$$

This notation is also extended to allow $X \rhd_s X$ for all X.

Lemma 1 (Basic properties of substitution).

(a) Every variable free in $[N/x]M$ is free in either N or M.

(b) If v is not free in M, then

$$[N/v][v/x]M \rhd_\alpha [N/x]M .$$

(c) If $v \not\equiv x$ and v is not free in N, then

$$[N/x][Z/v]M \rhd_\alpha [([N/x]Z)/v][N/x]M .$$

(d) Without any restriction,

$$[N/v][Z/v]M \rhd_\alpha [([N/v]Z)/v]M .$$

Proof. By induction on the number of occurrences of atoms in M. Part (d) can also be deduced from (b) and (c), as follows:

$$\begin{aligned}
[N/v][Z/v]M &\rhd_\alpha [N/v][Z/w][w/v]M \text{ by (b), (w is a new variable)}\\
&\rhd_\alpha [([N/v]Z)/w][N/v][w/v]M \text{ by (c),}\\
&\equiv [([N/v]Z)/w][w/v]M \text{ since } v \notin [w/v]M\\
&\rhd_\alpha [([N/v]Z)/v]M \text{ by (b).}
\end{aligned}$$

Parts (b) and (c) are also proved in [CLg. I] p. 95, Theorem 2(c), and (d) is proved on p. 103. (Incidentally, in (b), if no variable free in vN is bound in M, we can replace '$\rhd_\alpha$' by '≡'; similarly in (c) and (d) if no variable free in NZ is bound in M.)

Corollary 1.1. If U $\rhd_\beta$ V, then every variable free in V is free in U.

Proof. By (a).

The next two lemmas say in effect that congruent terms can be treated as identical.

Lemma 2 (Basic properties of congruence).

(a) The relation $\rhd_\alpha$ (called congruence) is transitive, reflexive and symmetric.

(b) If U $\rhd_\alpha$ V, then U and V contain exactly the same free variables.

(c) If N $\rhd_\alpha$ N' and M $\rhd_\alpha$ M', then

$$[N/x]M \rhd_\alpha [N'/x]M'.$$

Proof. Parts (a) and (b) are easy. For (c), use induction on the number of occurrences of atoms in M.

Lemma 3 (Congruence and β-reduction). If X $\rhd_\alpha$ Y $\rhd_\beta$ Z, then the β-contractions can be moved to the start of the reduction; i.e. there is a T such that

$$X \rhd_\beta T \rhd_\alpha Z .$$

Proof. It is enough to prove the result when the two given reductions are both single contractions, and show that in this case $X \rhd_\beta T$ turns out to be a single contraction. Suppose $X \rhd_\alpha Y$ by the contraction

$$\lambda u. W \rhd_\alpha \lambda v. [v/u]W ,$$

and suppose $Y \rhd_\beta Z$ by the contraction

$$(\lambda y. M)N \rhd_\beta [N/y]M .$$

Case 1: in Y, the contracted $(\lambda y. M)N$ does not overlap with $\lambda v. [v/u]W$. In this case there is an occurrence of $(\lambda y. M)N$ in X, not overlapping with $\lambda u. W$; and contracting $(\lambda y. M)N$ first, then $\lambda u. W$, gives Z.

Case 2: in Y, $(\lambda v. [v/u]W)$ is in M or in N. Then in X there is a β-redex of form $(\lambda y. M_0)N_0$, with

$$N_0 \rhd_\alpha N, \qquad M_0 \rhd_\alpha M .$$

Contracting this redex gives

$$[N_0/y]M_0 ,$$

which by Lemma 2(c), is congruent to $[N/y]M$.

Case 3: in Y, $\lambda v. [v/u]W$ is the same as $\lambda y. M$. Then in X there is a β-redex $(\lambda u. W)N$. Contracting this redex gives

$$[N/u]W .$$

By Lemma 1(b), since v is not free in W by the restriction on α-contractions, [N/u]W is congruent to

$$[N/v][v/u]W ,$$

which is [N/y]M, as required.

Case 4: in Y, $(\lambda y. M)N$ is in [v/u]W. Then X contains a redex $(\lambda x. M_0)N_0$ in the corresponding position in W, and

$$\begin{aligned} (\lambda y. M)N &\equiv [v/u]((\lambda x. M_0)N_0) \\ (1) \qquad &\equiv (\lambda x. [v/u]M_0)[v/u]N_0 , \end{aligned}$$

since $u \not\equiv x$ and $v \not\equiv x$. (This is because neither u nor v can be bound in W, by the definition of α-redex. For the same reason, the substitution [v/u] does not use clause (iv) or the second part of (v) in Definition 1. 4.)

Thus in X, we have

$$\lambda u. W ,$$

containing $(\lambda x. M_0)N_0$. Let T be the result of contracting this β-redex in X. Then in T we have a term

$$\lambda u. W^* ,$$

containing $[N_0/x]M_0$. Now the second part of Definition 1. 4(v) might be used in $[N_0/x]M_0$, and might introduce bound occurrences of v or u. But this can only happen if some free variables of N_0 occur bound in M_0. Let M_0' be the result of replacing any such variables by new ones. Then by Lemma 2(c),

$$[N_0/x]M_0 \triangleright_\alpha [N_0/x]M_0' .$$

This α-reduction will change $\lambda u. W^*$ in T to a new term

$$\lambda u. W' ,$$

which does not contain u or v bound. We next α-contract this to

$$\lambda v.[v/u]W',$$

containing $[v/u][N_0/x]M'_0$. And furthermore

$$\begin{aligned} [v/u][N_0/x]M'_0 &\rhd_\alpha [([v/u]N_0)/x][v/u]M'_0 && \text{by Lemma 1(c)} \\ &\rhd_\alpha [([v/u]N_0)/x][v/u]M_0 && \text{by Lemma 2(c)} \\ &\equiv [N/y]M \quad \text{by (1)}. \end{aligned}$$

Therefore we can α-reduce T to Z, as required.

Corollary 3.1. If $X \rhd_\alpha Y \rhd_s Z$, then $X \rhd_s Z$.

This completes the basic lemmas. The proof proper will be carried out in three main steps.

Step 1. If $M \rhd_s M'$ and $N \rhd_s N'$, then

$[N/x]M \rhd_s [N'/x]M'$.

Proof. By Lemmas 2 and 3, it is enough to prove the result under the assumption that no variables free in N are bound in M, and the given special reductions contain no α-contractions. This restricted result will now be proved by induction on M. (Compare Definition 1.4.)

Case 1: $M \equiv x$. Then $M' \equiv x$, and

$[N/x]M \equiv N \rhd_s N' \equiv [N'/x]M'$.

Case 2: $M \equiv a$, an atom distinct from x. Then $M' \equiv a$, and

$[N/x]M \equiv a \rhd_s a \equiv [N'/x]M'$.

Case 3: $M \equiv M_1 M_2$. If M is not one of the β-redexes $R_1, \ldots, R_n$ involved in the special reduction $M \rhd_s M'$, then each

R_i is in M_1 or M_2. Hence the reduction of M has form

$$\begin{aligned} M \equiv M_1 M_2 &\rhd_s M_1' M_2' \quad \text{by reductions in } M_1, M_2 \\ &\equiv M' . \end{aligned}$$

Then

$$\begin{aligned} [N/x]M &\equiv ([N/x]M_1)([N/x]M_2) \\ &\rhd_s ([N'/x]M_1')([N'/x]M_2') \quad \text{by induction hypothesis} \\ &\equiv [N'/x]M' . \end{aligned}$$

On the other hand, if M is an R_i, then the reduction of M has form

$$\begin{aligned} M \equiv R_i &\equiv (\lambda v.\, Y)Z \\ &\rhd_s (\lambda v.\, Y')Z' \quad \text{by reductions in } Y, Z \\ &\rhd [Z'/v]Y' \quad \text{by residual of } R_i \\ &\equiv M' . \end{aligned}$$

(The residual of R_i must be the last one to be contracted in the special reduction, because R_i contains the other redexes.)

In the case that $v \not\equiv x$, we note that v is not free in N by the restriction on bound variables of M; and then we have

$$\begin{aligned} [N/x]M &\equiv (\lambda v.\, [N/x]Y)([N/x]Z) \quad \text{since } v \notin N \\ &\rhd_s (\lambda v.\, [N'/x]Y')([N'/x]Z') \quad \text{by induction hypothesis} \\ &\rhd [([N'/x]Z')/v][N'/x]Y' \\ &\rhd_\alpha [N'/x][Z'/v]Y' \quad \text{by Lemmas 1(c) and 2(a)} \\ &\equiv [N'/x]M' . \end{aligned}$$

And by Lemma 3, this reduction can be made into a special one.

In the case that $v \equiv x$, we have

$$\begin{aligned} [N/x]M &\equiv (\lambda v.\, Y)([N/v]Z) \\ &\rhd_s (\lambda v.\, Y')([N'/v]Z') \quad \text{by induction hypothesis} \\ &\rhd [([N'/v]Z')/v]Y' \end{aligned}$$

$\triangleright_\alpha$ $[N'/v][Z'/v]Y'$ by Lemma 1(d)
$\equiv$ $[N'/x]M'$.

And by Lemma 3 this reduction can be made into a special one.

Case 4: $M \equiv \lambda y. M_1$ and $y \not\equiv x$. By the restriction on variables bound in M, y is not free in N; hence y is not free in N'. Also all β-redexes in M are in M_1, so M' has form $\lambda y. M_1'$, where $M_1 \triangleright_s M_1'$. Hence

$$
\begin{aligned}
[N/x]M &\equiv \lambda y. [N/x]M_1 && \text{since } y \notin N \\
&\triangleright_s \lambda y. [N'/x]M_1' && \text{by induction hypothesis} \\
&\equiv [N'/x]M' .
\end{aligned}
$$

Case 5: $M \equiv \lambda x. M_1$. Then M' has form $\lambda x. M_1'$, where $M_1 \triangleright_s M_1'$, and so we have

$$[N/x]M \equiv M \triangleright_s M' \equiv [N'/x]M' .$$

This completes Step 1. The next step is the key to the proof.

Step 2. If $U \triangleright_s X$ and $U \triangleright_s Y$, then there is a Z such that $X \triangleright_s Z$ and $Y \triangleright_s Z$.

Proof. By Lemma 3, it is enough to prove the result for the case that the given reductions contain no α-contractions. We shall use induction on U.

Case 1: U is an atom. Then $X \equiv Y \equiv U$. Choose $Z \equiv U$.

Case 2: $U \equiv \lambda x. U_1$. Then all β-redexes in U are in U_1, so we have

$$X \equiv \lambda x. X_1, \qquad Y \equiv \lambda x. Y_1 ,$$

where $U_1 \triangleright_s X_1$ and $U_1 \triangleright_s Y_1$. Hence, by the induction hypothesis there is a Z_1 such that

$$X_1 \triangleright_s Z_1 , \qquad Y_1 \triangleright_s Z_1 .$$

Choose $Z \equiv \lambda x. Z_1$.

<u>Case 3:</u> $U \equiv U_1 U_2$. If the β-redexes $R_1, \ldots, R_n$ for both the given special reductions are in U_1 and U_2, then

$$X \equiv X_1 X_2, \qquad Y \equiv Y_1 Y_2 ,$$

where $U_i \triangleright_s X_i$ and $U_i \triangleright_s Y_i$ for $i = 1, 2$. Then by the induction hypothesis there are Z_1, Z_2 such that $X_i \triangleright_s Z_i$ and $Y_i \triangleright_s Z_i$. Choose $Z \equiv Z_1 Z_2$.

Now suppose that for the reduction $U \triangleright_s X$, one R_j is U itself. Then the reduction has form

$$\begin{aligned} U \equiv R_j &\equiv (\lambda x. M)N \\ &\triangleright_s (\lambda x. M^*)N^* \quad \text{by reduction in } M, N \\ &\triangleright [N^*/x]M^* \\ &\equiv X . \end{aligned}$$

If the reduction $U \triangleright_s Y$ does not involve contracting the residual of U, then it has form

$$\begin{aligned} U &\equiv (\lambda x. M)N \\ &\triangleright_s (\lambda x. M')N' \\ &\equiv Y . \end{aligned}$$

Then the induction hypothesis gives us M_1 and N_1 such that

(2) $\qquad M^* \triangleright_s M_1, \qquad M' \triangleright_s M_1,$

(3) $\qquad N^* \triangleright_s N_1 , \qquad N' \triangleright_s N_1 .$

Choose $Z \equiv [N_1/x]M_1$. Then we have

$$\begin{aligned} X &\equiv [N^*/x]M^* \\ &\triangleright_s Z \qquad \text{by Step 1} , \end{aligned}$$

and

$$Y \equiv (\lambda x.\, M')N' \;\triangleright_s\; (\lambda x.\, M_1)N_1$$
$$\triangleright\; [N_1/x]M_1 \;\equiv\; Z\,.$$

And by Lemma 3, we can make this last reduction into a special one.

Finally, if the reduction $U \triangleright_s Y$ does involve contracting the residual of U, then it must have form

$$\begin{aligned} U &\equiv (\lambda x.\, M)N \\ &\triangleright_s (\lambda x.\, M')N' \\ &\triangleright [N'/x]M' \\ &\equiv Y\,. \end{aligned}$$

Then the induction hypothesis gives us M_1 and N_1 satisfying (2) and (3). Choose Z as above. This completes Step 2.

Step 3. <u>If</u> $U \triangleright Y$ <u>and</u> $U \triangleright X$, <u>then there is a</u> Z <u>such that</u> $X \triangleright Z$ <u>and</u> $Y \triangleright Z$.

Proof. (Recall that '$\triangleright$' denotes any reduction by β- and α-contractions.) An induction on the number of contractions in the reduction from U to X (see Figure 1) shows that it is enough to prove

$$\text{(4)}\qquad \begin{aligned} &U \text{ contracts to } X \text{ and } U \triangleright Y \\ &\Rightarrow \exists Z\colon X \triangleright Z \text{ and } Y \triangleright Z\,. \end{aligned}$$

If the contraction in (4) is an α-contraction, then it can be reversed, giving $X \triangleright U \triangleright X$. So we can choose $Z \equiv Y$ in this case. If the contraction is a β-contraction, then it is a one-step special reduction. Hence (4) follows from

$$\text{(5)}\qquad \begin{aligned} &U \triangleright_s X \text{ and } U \triangleright Y \\ &\Rightarrow \exists Z\colon X \triangleright Z \text{ and } Y \triangleright_s Z\,. \end{aligned}$$

But (5) follows from Step 2 and Lemma 3, by induction on the number of contractions from U to Y. (See Figure 2.)

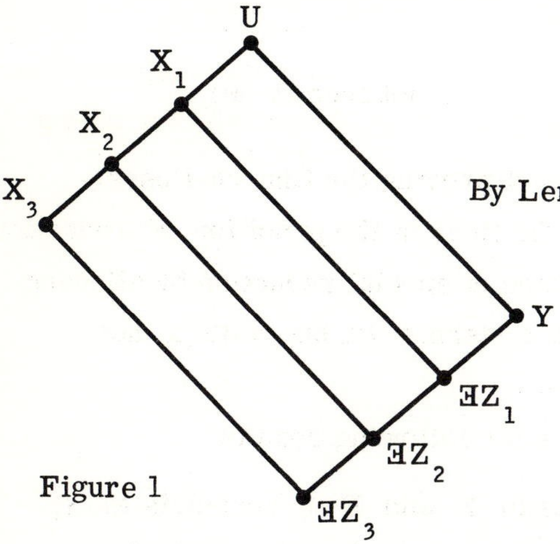

Figure 1

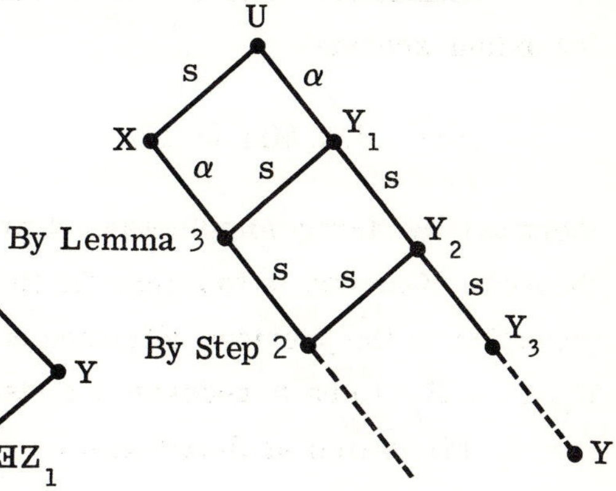

Figure 2

Remark 1. The Church-Rosser theorem for combinatory weak reduction can also be proved by the above method. Redexes would be terms of form

$$\mathbf{I}X, \quad \mathbf{K}XY, \quad \mathbf{S}XYZ\,.$$

A special reduction would be defined as a simultaneous contraction of non-overlapping redexes. With this simplification, Step 2 would be easier than for λ-reduction, Step 1 and Lemmas 2 and 3 would be redundant, and Step 3 would be the same. Essentially similar proofs are in Rosser [MLV] and Sanchis [FDR]. (Incidentally, if we tried to simplify special λ-reductions in the same way, Step 2 would fail.)

This method still works when we modify the definition of combinatory reduction by adding axiom-schemes

$$(3.11) \qquad \mathbf{R}XY\bar{0} \rhd X, \quad \mathbf{R}XY(\overline{k+1}) \rhd Y\bar{k}(\mathbf{R}XY\bar{k})\,,$$

or axiom-schemes

(3.12) $\mathbf{Z}\bar{n} \triangleright \mathbf{Z}_n$.

Remark 2. For strong λ-reduction (Definition 7.3) with the axiom-scheme

$$(\eta) \qquad (\lambda x.\, Mx) \triangleright M \qquad (x \text{ not free in } M),$$

there are two fairly simple ways of proving the Church-Rosser theorem. One way is to follow the lines of the proof for $\alpha\beta$-reduction, generalising the previous definition of special reduction by allowing $R_1, \ldots, R_n$ to be η-redexes (i. e. terms $\lambda x.\, Mx$ with x not free in M) as well as β-redexes.

The other way is to prove the following results:

(6) If U η-contracts to X and U η-contracts to Y, then either $X \equiv Y$ or there is a Z such that X η-contracts to Z and Y η-contracts to Z; (see Figure 3).

(7) If U η-contracts to X and U' is congruent to U, then U' η-contracts to an X' congruent to X.

(8) If U η-contracts to X and U β-contracts to Y, then either X is congruent to Y or there is a Z such that X β-contracts to Z and Y η-reduces to Z. (See Figure 4.)

From (6) one can deduce the Church-Rosser theorem for $\triangleright_\eta$ as in Step 3 above. Then from (8), (7) and the theorem for $\triangleright_{\alpha\beta}$, one can deduce the theorem for $\triangleright_{\alpha\beta\eta}$ by easy inductions.

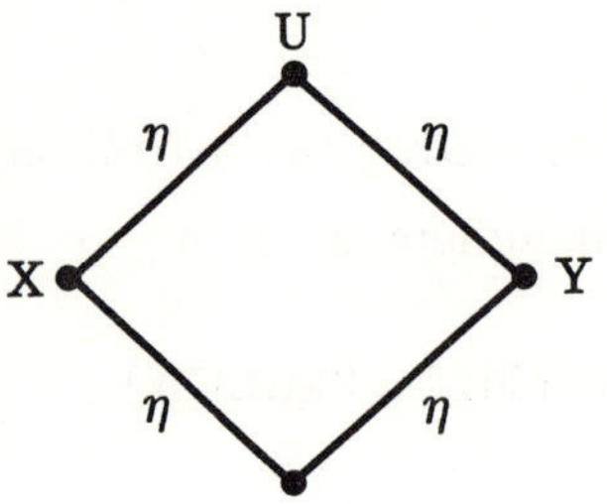

Figure 3

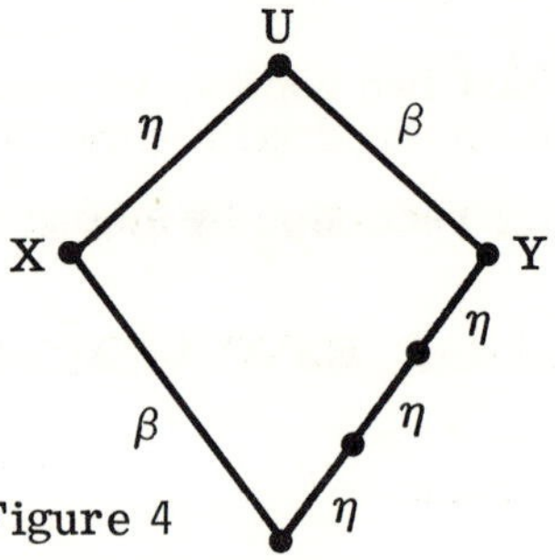

Figure 4

Appendix 2
Proof of Theorem 11.17

Theorem 11.17 says that if $\mathscr{F}$ is provable in arithmetic, then $\mathscr{F}^*$ is true. The proof is by induction on the proof of $\mathscr{F}$, and is divided into cases according as $\mathscr{F}$ is an axiom or the conclusion of a rule. As examples we shall consider here one axiom-scheme, the first in Definition 11.10, and the most difficult deduction-rule, that of induction. The other axiom-schemes and rules will be left as a tedious but not very difficult exercise.

For the first axiom-scheme, suppose that $\mathscr{F}$ has the form

$$\mathscr{A} \supset (\mathscr{A} \,\&\, \mathscr{A}),$$

and for simplicity suppose that $\mathscr{A}^*$ contains only three variables, one free and one bound by each quantifier. Then in the shortened notation of Definition 11.14,

$$\mathscr{A}^* \equiv \exists y \forall z\, A\{y, z\}.$$

To construct $\mathscr{F}^*$, we start with three 'copies' of $\mathscr{A}^*$ with distinct quantified variables (but the same free variable):

$$\exists y_1 \forall z_1 A\{y_1, z_1\}, \quad \exists y_2 \forall z_2 A\{y_2, z_2\}, \quad \exists y_3 \forall z_3 A\{y_3, z_3\}$$

Using the last two copies, we construct

$$(\mathscr{A} \,\&\, \mathscr{A})^* \equiv \exists y_2, y_3 \forall z_2, z_3 (A\{y_2, z_2\} \,\&\, A\{y_3, z_3\})$$

Then by the analogue of Definition 11.14(iid) for the case $q \equiv p \equiv 2$, $\mathscr{F}^*$ is

$$\exists z_1', y_2', y_3' \forall y_1, z_2, z_3 \left(\begin{array}{c} A\{y_1, z_1'y_1z_2z_3\} \\ \overline{\supset} \\ (A\{y_2'y_1, z_2\} \mathbin{\overline{\&}} A\{y_3'y_1, z_3\}) \end{array} \right) .$$

Let x be the free variable in $\mathscr{F}^*$ corresponding to the one free in $\mathscr{A}$. To show that $\mathscr{F}^*$ is true, define

$$Y_2' \equiv [x, y_1]y_1 ,$$

$$Y_3' \equiv [x, y_1]y_1 ,$$

$$Z_1' \equiv [x, y_1, z_2, z_3]\mathbf{D}z_2z_3A_3 ,$$

where $A_3 \equiv A\{y_1, z_3\}$ and **D** is from Lemma 11.9. Then when $Y_2'x$, $Y_3'x$, $Z_1'x$ are substituted for y_2', y_3', z_1' in the matrix of $\mathscr{F}^*$, the result can be reduced to

$$A\{y_1, \mathbf{D}z_2z_3A_3\} \overline{\supset} (A\{y_1, z_2\} \mathbin{\overline{\&}} A_3) .$$

Now when any constant terms **X**, Y_1, Z_2, Z_3 are substituted for x, y_1, z_2, z_3 in this term, the result will have form

$$A\{Y_1, \mathbf{D}Z_2Z_3A_3'\} \overline{\supset} (A\{Y_1, Z_2\} \mathbin{\overline{\&}} A_3')$$

where A_3' is a constant with type **N** and hence reduces to a numeral. Also $A\{Y_1, Z_2\}$ reduces to a numeral.

<u>Case 1</u>: $A_3' \rhd \overline{0}$. Then by Lemma 11.9,

$$\mathbf{D}Z_2Z_3A_3' \rhd Z_2 ,$$

and hence the whole term above reduces to

$$A\{Y_1, Z_2\} \overline{\supset} (A\{Y_1, Z_2\} \mathbin{\overline{\&}} \overline{0}) ,$$

which reduces to $\overline{0}$ by definition of $\overline{\supset}$, $\overline{\&}$.

Case 2: $A'_3 \triangleright \overline{k+1}$ for some k. Then by Lemma 11.9,

$$DZ_2Z_3A'_3 \triangleright Z_3 ,$$

and hence the whole term above reduces to

$$A'_3 \mathbin{\bar{\supset}} (A\{Y_1, Z_2\} \mathbin{\bar{\&}} \overline{k+1}) ,$$

which reduces further to

$$\overline{k+1} \mathbin{\bar{\supset}} (A\{Y_1, Z_2\} \mathbin{\bar{\&}} \overline{k+1}) ,$$

which reduces to $\overline{0}$ by definition of $\bar{\supset}$.

Thus if $\mathscr{F}$ is an instance of the first axiom-scheme, $\mathscr{F}^*$ is true.

The most difficult deduction-rule to deal with is that for induction:

$$\frac{\mathscr{A}(0) \qquad \forall x(\mathscr{A}(x) \supset \mathscr{A}(x'))}{\mathscr{A}(x) .}$$

Suppose for simplicity that $\mathscr{A}(x)^*$ has the form

(1) $\quad \mathscr{A}(x)^* \equiv \exists y \forall z A\{x^*, y, z\} .$

Then by induction on $\mathscr{A}(x)$ it can easily be shown that

(2) $\quad \mathscr{A}(0)^* \equiv \exists y \forall z A\{\overline{0}, y, z\},$

$\quad \mathscr{A}(x')^* \equiv \exists y \forall z A\{\bar{\sigma}x^*, y, z\} .$

To construct $(\forall x(\mathscr{A}(x) \supset \mathscr{A}(x')))^*$, we first take two 'copies',

$$\exists y_1 \forall z_1 A\{x^*, y_1, z_1\}, \qquad \exists y_2 \forall z_2 A\{\bar{\sigma}x^*, y_2, z_2\} ,$$

and then apply Definition 11.14(iid) to get

$$\exists z'_1, y'_2 \ \forall y_1, z_2 \ (A\{x^*, y_1, z'_1 y_1 z_2\} \supset A\{\bar{\sigma}x^*, y'_2 y_1, z_2\}).$$

Finally by (iif) we have

$$(3) \quad (\forall x(\mathscr{A}(x) \supset \mathscr{A}(x')))^* \equiv \exists z''_1, y''_2 \ \forall x^*, y_1, z_2 \begin{pmatrix} A\{x^*, y_1, z''_1 x^* y_1 z_2\} \\ \supset \\ A\{\bar{\sigma}x^*, y''_2 x^* y_1, z_2\} \end{pmatrix}$$

To deal with the induction-rule, we must show that if the formulae (2) and (3) are true, then (1) is true. Therefore suppose that (2) and (3) are true, and for simplicity suppose that there is only one free variable in $\mathscr{A}(0)$; let v be the corresponding variable in (1), (2) and (3). Then there exist constant terms Y, Z_1, Y_2 such that the two formulae

$$(4) \quad \begin{array}{l} A\{\bar{0}, Yv, z\} \ \rhd \bar{0}, \\ (A\{x^*, y_1, Z_1 v x^* y_1 z_2\} \supset A\{\bar{\sigma}x^*, Y_2 v x^* y_1 z_2\}) \rhd \bar{0} \end{array}$$

are true no matter what constants are substituted for their variables.

To show that (1) is true, we first define $Y' \equiv [v]\mathbf{R}(Yv)(Y_2 v)$. Then Y' is a constant term with the property that

$$(5) \quad \begin{array}{ll} Y'v\bar{0} = Yv, & \\ Y'v\overline{(m+1)} = Y_2 v\bar{m}(Y'v\bar{m}) & \text{for all } m. \end{array}$$

Then substituting $Y'vx^*$ into the matrix of (1) gives

$$A\{x^*, Y'vx^*, z\}.$$

Since x^* has type **N**, any constant substituted for it must reduce to a numeral; hence to prove (1) true, it is enough to show that for all k and all constant V and Z, the term

$$(6) \quad A\{\bar{k}, Y'V\bar{k}, Z\}$$

reduces to $\bar{0}$. This is done by induction on k, as follows.

If $k \equiv 0$: then by (5), $Y'V\bar{0} = YV$, so by (4),

$$A\{\bar{k},\ Y'V\bar{0},\ Z\} = \bar{0}\ .$$

Since $\bar{0}$ is irreducible, this '$= \bar{0}$' implies '$\triangleright \bar{0}$'.

If $k \equiv m + 1$: then by (5), the term (6) is equal to

$$(7) \qquad A\{\overline{m+1},\ Y_2V\bar{m}(Y'V\bar{m}),\ Z\}\ .$$

Now by the second part of (4) with $\bar{m}$ for x^*, $Y'V\bar{m}$ for y_1, and Z for z_2, the term

$$A\{\bar{m},\ Y'V\bar{m},\ W\} \mathbin{\bar{\supset}} A\{\overline{m+1},\ Y_2V\bar{m}(Y'V\bar{m}),\ Z\}$$

reduces to $\bar{0}$. (Here W is $Z_1V\bar{m}(Y'V\bar{m})Z$.) But the first half of this term is a special case of (6) with $\bar{m}$ for $\bar{k}$ and W instead of Z. Therefore by the induction-hypothesis, this term reduces to $\bar{0}$. Hence by the definition of $\bar{\supset}$, the second half of the above term must reduce to $\bar{0}$ too. But this second half is (7), to which (6) is equal. Therefore (6) reduces to $\bar{0}$, completing the induction-step.

This completes the treatment of the induction-rule, in the case that $\mathscr{A}^*$ has only two quantified variables.

Unfortunately the general case is not so simple. Suppose now that instead of (1) we have

$$\mathscr{A}(x)^* \equiv \exists y_1, \ldots, y_n \forall z\, A\{x^*, y_1, \ldots, y_n, z\}\ ,$$

where $n \geq 2$ and $y_1, \ldots, y_n$ have types $\alpha_1, \ldots, \alpha_n$ respectively. For simplicity suppose that x is the only free variable in $\mathscr{A}(x)^*$. Then the previous argument will work if we can find constants $Y'_1, \ldots, Y'_n$ such that for each $i = 1, \ldots, n$,

$$(8) \quad \begin{aligned} Y'_i\overline{0} &= A_i , \\ Y'_i(\overline{m+1}) &= B_i\overline{m}(Y'_1\overline{m})\ldots(Y'_n\overline{m}) , \end{aligned}$$

where $A_1, \ldots, A_n$ are the n analogues of the Y in (5), and $B_1, \ldots, B_n$ are the analogues of Y_2, and the v in (5) does not now occur.

If $\alpha_1, \ldots, \alpha_n$ are identical, suitable $Y'_1, \ldots, Y'_n$ can be constructed from the **D**'s in Lemma 11.9, but otherwise we can use the following construction, due to K. Schütte. It proceeds by induction on n.

For $n = 1$: define $Y'_1 \equiv \mathbf{R}A_1B_1$.

For $n > 1$: first define

$$C \equiv [u_1, \ldots, u_{n-1}]\mathbf{R}A_n([x]B_nx(u_1x)\ldots(u_{n-1}x)) ,$$

$$E_i \equiv [x, u_1, \ldots, u_{n-1}, w]\mathbf{D}(u_iw)(B_ix(u_1x)\ldots(u_{n-1}x)(Cu_1\ldots u_{n-1}x))$$

for $i = 1, \ldots, n-1$. (**D** comes from Lemma 11.9.) Then by the induction-hypothesis we can construct terms $G_1, \ldots, G_{n-1}$ such that

$$(9) \quad \begin{aligned} G_i\overline{0} &= \mathbf{K}A_i , \\ G_i(\overline{m+1}) &= E_i\overline{m}(G_1\overline{m})\ldots(G_{n-1}\overline{m}) . \end{aligned}$$

Using these, we define for $i = 1, \ldots, n-1$,

$$Y'_i \equiv [x]G_ixx ,$$

and define

$$Y'_n \equiv [x]C(G_1x)\ldots(G_{n-1}x)x .$$

The proof that these satisfy (8) is fairly straightforward.

Bibliography

Andrews, P. B.

[TTT] A Transfinite Type Theory with Type Variables. North-Holland Co., 1965.

Barendregt, H. P.

[SEM] Some Extensional Term Models for Combinatory Logics and λ-calculi. Thesis, University of Utrecht, 1971.

Böhm, C.

[PFB] Alcune proprieta della forme β-η-normali del λ-K-calcolo. Publicazioni dell'instituto per le applicazioni del calcolo (Consiglio nazionale delle ricerce Roma), no. 696 (1968).

Böhm, C. and Gross, W.

[ICC] Introduction to the CUCH. In 'Automata Theory', Academic Press, 1966.

Bunder, M. V. W.

[STB] Set Theory based on Combinatory Logic. Doctoral thesis, University of Amsterdam, 1969.

Church, A.

[CLC] The calculi of Lambda-conversion. Princeton University Press, 1941.

[FST] A formalization of the simple theory of types. Journal of Symbolic Logic 5 (1940), pp. 56-68.

[NEP] A note on the Entscheidungsproblem. Journal of Symbolic Logic 1 (1936), pp. 40-1, 101-2.

[UPE] An unsolvable problem of elementary number theory. American Journal of Mathematics 58 (1936), pp.345-63.

Curry, H. B.

[CLg. I] (with Feys, R. and Craig, W.) Combinatory logic, volume I. North-Holland Co., 1958.

[CLg. II] (with Seldin, J. and Hindley, R.) Combinatory logic, volume II. North-Holland Co., 1971.

[FML] Foundations of Mathematical Logic. McGraw-Hill Co., 1963.

[IFL] The inconsistency of certain formal logics. Journal of Symbolic Logic 7 (1942), pp. 115-17.

[MBF] Modified basic functionality in combinatory logic. Dialectica 23 (1969), pp. 83-92.

Fitch, F. B.

[CCM] A complete and consistent modal set theory. Journal of Symbolic Logic 32 (1967), pp. 93-103.

[SCD] The system CΔ of combinatory logic. Journal of Symbolic Logic 28 (1963), pp. 87-97.

[SFL] A system of formal logic without an analogue to the Curry W operator. Journal of Symbolic Logic 1 (1936), pp. 92-100.

Gentzen, G.

[ILD] Investigations into Logical Deduction. In 'The Collected Papers of Gerhard Gentzen', North-Holland Co., 1969.

Gödel, K.

[BNN] Über eine bisher noch nicht benutzte Erweiterung des finiten Standpunktes. Dialectica 12 (1958), pp. 280-7.

Goodman, N.

[IAT] Intuitionistic arithmetic as a theory of constructions. Doctoral thesis, Stanford University, USA, 1968. (See 'Intuitionism and Proof Theory', North-Holland Co., 1970.)

Grzegorczyk, A.

[ROA] Recursive objects in all finite types. Fundamenta Mathematicae 54 (1964), pp. 73-93.

Henkin, L.

[CTT] Completeness in the theory of types. Journal of Symbolic Logic 15 (1950), pp. 81-91.

[TPT] A theory of propositional types. Fundamenta Mathematicae 52 (1963), pp. 323-44.

Hindley, J. R.

[ASR] Axioms for strong reduction in combinatory logic. Journal of Symbolic Logic 32 (1967), pp. 224-36.

[PTO] The principal type-scheme of an object in combinatory logic. Transactions of the American Mathematical Society 146 (1969), pp. 29-60.

Hindley, J. R. and Lercher, B.

[SPC] A short proof of Curry's normal form theorem. Proceedings of the American Mathematical Society 24 (1970), pp. 808-10.

Kleene, S. C.

[IMM] Introduction to metamathematics. Van Nostrand Co., Noordhoff Co., 1952.

[LDR] λ-definability and recursiveness. Duke Math. Journal 2 (1936), pp. 340-53.

Landin, P. J.

[MEE] The mechanical evaluation of expressions. Computer Journal January 1964, pp. 308-20.

[CAC] A correspondence between ALGOL-60 and Church's λ-notation. Communications of the Association for Computing Machinery 8 (1965), pp. 89-101, 158-65.

Lercher, B.

[SRN] Strong reduction and normal form in combinatory logic. Journal of Symbolic Logic 32 (1967), pp. 213-23.

[DHA] The decidability of Hindley's axioms for strong reduction. Journal of Symbolic Logic 32 (1967), pp. 237-9.

Orgass, R. J.

[SRP] Some results concerning proofs of statements about programs. Journal of Computer and Systems Sciences 4 (1970), pp. 74-88.

Orgass, R. J. and Fitch, F. B.

[TCM] A theory of computing machines. Studium Generale 22 (1969), pp. 83-104, 113-36.

Prawitz, D.

[HHO] Hauptsatz for higher order logic. Journal for Symbolic Logic 33 (1968), pp. 452-7.

Rosser, J. B.

[MLV] A mathematical logic without variables. Annals of Maths. (2) 36 (1935), pp. 127-50, and Duke Math. Journal 1 (1935), pp. 328-55.

Sanchis, L. E.

[FDR] Functionals defined by recursion. Notre Dame Journal of Formal Logic 8 (1967), pp. 161-74.

[TCL] Types of combinatory logic. Notre Dame Journal of Formal Logic 5 (1964), pp. 161-80.

Schönfinkel, M.

[BML] Über die Bausteine der mathematischen Logik. Math. Annalen 92 (1924), pp. 305-16; English translation in 'From Frege to Gödel; a source-book in mathematical logic', Harvard University Press, 1967.

Schütte, K.

[SSP] Syntactical and semantical properties of simple type theory. Journal of Symbolic Logic 25 (1960), pp. 305-26.

[STT] On Simple Type Theory with Extensionality. In 'Logic, Methodology and Philosophy of Science III', North-Holland Co., 1968.

[TFE] Theorie der Funktionale endlicher Typen. Mimeographed notes, Munich University, 1968.

Scott, D. S.

[LTM] A lattice theoretic model for the λ-calculus. To be published.

[OMT] Outline of a Mathematical Theory of Computation. In Proceedings of the fourth annual Princeton conference on information sciences and systems (1970) pp. 196-176.

Seldin, J. P.

[SIC] Studies in illative combinatory logic. Doctoral thesis, University of Amsterdam, 1968.

Spector, C.

[PRA] Provably Recursive Functionals of Analysis. In 'Proceedings of Symposia in Pure Mathematics, V', American Mathematical Society 1962.

Strachey, C.

[FCP] Fundamental Concepts in Programming Languages. NATO conference, Copenhagen, 1967.

[TFS] Towards a Formal Semantics. In 'Formal description Languages for Computer Programming', North-Holland Co., 1965.

Strong, H. R.

[AGR] Algebraically generalized recursive function theory. IBM Journal of Research and Development, November 1968, pp. 465-75.

Tait, W. W.

[IIF] Intensional interpretations of functionals of finite type. Journal of Symbolic Logic 32 (1967), pp. 198-212.

[ILT] Infinitely Long Terms of Transfinite Type. In 'Formal Systems and Recursive Functions', North-Holland Co., 1965.

Takahashi, M.

[PCE] A proof of cut-elimination theorem in simple type-theory. Journal of the Mathematical Society of Japan 19 (1967), pp. 399-410.

Wagner, E.

[URS] Uniformly reflexive structures. Transactions of the American Math. Society 144 (1969), pp. 1-41.

Index

Greek letters and other symbols are listed at the end.

MISCELLANEOUS SYMBOLS

Main relation-symbols: